P9-EES-319

# Lives of the Planets

# LIVES
## *of the*
# PLANETS

*A Natural History of the Solar System*

RICHARD CORFIELD

A Member of the Perseus Books Group
New York

Published by Basic Books,
A Member of the Perseus Books Group



Books published by Basic Books are available at special discounts for bulk purchases in the United States by corporations, institutions, and other organizations. For more information, please contact the Special Markets Department at the Perseus Books Group, 11 Cambridge Center, Cambridge MA 02142, or call (617) 252–5298 or (800) 255–1514, or e-mail special.markets@perseusbooks.com.

*Interior design by Jeff Williams*

Library of Congress Cataloging-in-Publication Data
Corfield, R. M.
Lives of the planets : a natural history of the solar system / by Richard Corfield. — 1st ed.
p. cm.
Includes index.
ISBN-13: 978-0-465-01403-3 (hc : alk. paper)
ISBN-10: 0-465-01403-8 (hc : alk. paper) 1. Solar system. 2. Planetology. 3. Space probes. I. Title.

QB501.C6619 2007
523.2—dc22

2007013808

10 9 8 7 6 5 4 3 2 1

*For Susie*

Planets are like organisms in that they have irreducible individuality and must therefore be explained by methods of historical analysis: they are not like molecules in a chemical equation.

—STEPHEN JAY GOULD

Watching the delicate artistry with which spacecraft are fashioned, trembling at the drama of their launches, admiring the exquisite curves of their trajectories, listening to their strange voices speaking from millions of miles away, stretching the imagination to envisage the new concepts of space, you realise that this is . . . an unheard-of poetry of tremendous dimensions. And who questions the cost of a poem?

—ALFRED BESTER

# CONTENTS

# PREFACE

Just off Muswell Hill Broadway, in North London, is a narrow alley. The open stalls that once backed onto this rather ordinary North London street are now long gone, converted into apartments in an area much sought after by young professionals intent on pursuing village life while remaining within safe distance of a Starbucks. A few hundred feet to the northeast is Muswell Hill roundabout and, just down the hill, Alexandra Palace, the Victorian pile from which the BBC started broadcasting television in the days between the two World Wars.

But for me, this area has another, completely private claim on memory, for it was here in the early 1970s that I asked the local carpenter to build me a spacecraft.

The air was thick with the resinous scent of pine, and the youthful owner, clearly engrossed with the business of his trade, looked workmanlike in his brown leather apron. Yet he did not blink when I, with my father standing by, asked him to build me the nose cone assembly of a *Saturn V*. To an eight-year-old, it seemed the most reasonable request in the world; after all, how hard could it be? NASA was doing it all the time.

He listened gravely as I told him that all I required was the nose cone, because I had already made arrangements for the rocket's fuselage. Indeed, the old wooden coat rack I had salvaged was even at that moment waiting on the launching pad in our backyard. The carpenter did not laugh. Instead, he explained courteously that, although constructing the nose cone of a 364-foot-tall rocket that produced over 7 million pounds of thrust and was currently costing the American economy $3 billion a year was beyond his immediate art, he thought my answer might be to get hold of an old wooden barrel—say, an old beer barrel—and modify that.

Thanking him, my father and I (somewhat hurt) moved on.

For months afterward, the Americans' battle for the moon occupied my every waking moment. I remember, as though it were yesterday, the treat of staying up late on the night of July 20, 1969. The famous grainy images transmitted from the Sea of Tranquility were not helped by the technology—an old Bakelite television set in my grandfather's bedroom—available to the household of an impoverished Scottish minister in the late 1960s, but it sufficed.

At that moment—02:56:15 Universal Time (GMT), July 21, 1969—I was there, standing beside them, as Neil Armstrong and Buzz Aldrin made first contact with the surface of a celestial body that was not the planet on which we evolved. To this day, that memory still raises the hair on my arms. Our generation was *there*.

Forget Woodstock, forget the Berlin Wall, forget the Vietnam War. In a thousand years, the 1960s will be remembered for one thing only: the moment when 3.8 billion years of evolution on an unremarkable, watery world orbiting an unremarkable star in an unremarkable galaxy produced animals capable of traveling to other worlds. What is the statistical probability of being part of that one generation out of all the countless trillions since the first self-copying molecules aggregated on a layer of clay almost 4 billion years ago? Slim, I think you will agree.

As early as the late 1950s, unmanned missions were rocketing into space. Purely for the purposes of science, the Americans had planned to put a satellite into orbit as part of their contribution to the 1958 International Geophysical Year. Orderly progress, however, was derailed when, on October 4, 1957, the Soviets lofted *Sputnik 1*, a 59-cm aluminum sphere, into orbit. The Soviet craft—little more than a ball with a breadboarded radio beacon on board—caused mass hysteria in the United States because of the so-called "missile gap." The prevailing belief was that the Soviets would soon be dropping nuclear bombs on America like "rocks from a highway overpass" (as Lyndon Johnson famously put it) and that the American scientific and industrial establishment had very seriously "screwed the pooch" (as Chuck "Right Stuff" Yeager might say).

*Explorer 1* became America's first orbiting satellite after its successful launch on January 31, 1958. It returned instant results, catapulting the astronomer James Van Allen into the limelight with his discovery of the radiation belts that now bear his name. Space quickly became a political game, however, and the unmanned probes were relegated to insignificance for the next decade and a half as the Apollo program and the race for the moon dominated the public imagination. Space probes continued to be built, though, developing in sophistication as their designers strove to comprehend the scientific requirements of space exploration. It was a time of innovation and imagination, for the engineers and fabricators designing these unsung spacecraft had little idea of what they were trying to measure. What *was* important out there? Magnetism, surface chemistry, atmospheric composition, solar wind? All these properties, and more, would be addressed by the backroom boys of the space age, today allowing us to send unmanned robots equipped with picks, grinders, and lenses to the surface of Mars to carry out geology experiments, exactly as their human equivalents did nearly forty years ago on the lunar surface.

The story of the planets of our solar system is one of several interlocking pieces. The first piece is made up of all those individuals, from the ancient Babylonians to modern scientists, who devoted their lives to investigating those enigmatic bodies. The second consists of the space probes that, since I was young, have relayed information that made those planets real to us in a way impossible before the invention of today's rockets. The third piece is perhaps the grandest, since it relies on a synthesis of the first two: the natural history of the planets themselves. We now think of the planets as *places* rather than as dots in the sky and appreciate perhaps the greatest achievement yet in the new science of comparative planetology: the understanding that every body in the solar system has its own unique history. The planets and moons of the solar system were formed at the same time, approximately 4.6 billion years ago, but each of them, like each of us, has its own story, a composite of billions of contingent events. These stories are still mostly unknown; indeed, it would be presumptuous to say that we have more than scratched the surface of what can be discovered. But we are now at a stage where we can begin to understand the lives of the planets in the context of our own history of life on Earth.

The modern study of our solar system is where science and history intersect in a unique way. The *Voyager* and *Pioneer* deep-space probes, for example, left Earth decades ago and are still going strong, and who knows what they may yet discover. Simultaneously they are our past and our future.

RICHARD CORFIELD
*Long Hanborough, England*
2007

# 1

# Ages of Dark and Light: The Sun

Not far north of the British cathedral city of Salisbury is a sprawling area of chalky farmland. Its forbidding signs warn off the unwary traveler, and the locals do not visit; they know that this tranquil, quintessentially English countryside hides a deadly secret. Salisbury Plain is littered with unspent ammunition, tank traps, and other—unspecified but certainly nastier—paraphernalia of war.

South of this eerie place runs A303, the highway connecting the small town of Andover in the east with the market town of Ilminster, gateway to the moorlands of Devon and Cornwall in the west. Like Salisbury Plain itself, A303 is not for the fainthearted. Despite the many speed cameras that dot its sixty-mile length, it is popular with sports car drivers, for its long straightaways, sweeping curves, and two-lane width make it a fine place to put the Ferrari's hammer down.

The narrow strip of land sandwiched between this traffic black-spot and military proving ground is not the most obvious place to find one of mankind's oldest astronomical artifacts, but, nevertheless, it is there: a strange, horseshoe-shaped arrangement of giant standing stones measuring thirty meters in diameter—Stonehenge. If you stand at the center of the formation on June 21, the summer solstice, the sun rises precisely between the arms of the horseshoe.

## THE STONE COMPUTER

Stonehenge is a Stone Age supercomputer whose read-only memory consists of thirty-five megaliths, each weighing more than twenty-five tons. Its RAM is an enigmatic set of concentric holes in the ground, its monitor is a solitary megalith standing some distance apart from the others, and its hard drive is a 5 trillion-ton sphere that rotates once every twenty-four hours. This supercomputer's terrifying hydrogen fusion power supply is located 93 million miles from the rest of the hardware. For three thousand years, that supercomputer helped Neolithic peoples keep pace with the turning of the seasons, telling them when to sow and when to reap and predicting the occurrence of crucial astronomical events such as eclipses.

It was realized as long ago as the eighteenth century that Stonehenge's arrangement of stones was not likely to be accidental. William Stukeley, an antiquarian and cabalistic scholar with something of a dragon fixation, is generally credited with linking Stonehenge with a Neolithic sun cult. Over the next two centuries, as work proceeded on unraveling the mysteries of Stonehenge, it became apparent that the artifact was vastly more complicated than was originally thought.

We now know that Stonehenge was built over a fifteen hundred-year period from about 3100 BC to 1500 BC in a three-phase construction process. The first construction was a henge in the typical manner—a raised bank surrounded by a ditch. Nearly all traces of the second phase are gone, although it appears that a timber structure was built within the existing henge sometime in the third millennium BC, the early to middle Bronze Age. The famous trilithon stones were part of the final stage of construction. The first phase, although unremarkable now when compared to the giant trilithons and bluestones of the later period, is of most interest because it shows a knowledge of astronomy quite incredible for humans living centuries before the birth of Christ. On the northeast side of the henge, the circular bank and ditch are interrupted by an avenue leading off to the northeast. Slightly outside the henge's perimeter, nestling between the banks of this avenue, is the solitary heelstone. From the center of the circle, the sun rises exactly behind the heelstone on midsummer's day.

Dotted around the periphery of the henge, just inside the bank, are fifty-six holes discovered by John Aubrey, a seventeenth-century English antiquary and writer. The Aubrey Holes, as they are known, together with four large station stones that form an enigmatic rectangle superimposed on the periphery of the henge, date from the first phase of Stonehenge's construction. The significance of the Aubrey Holes and the station stones was a mystery well into the twentieth century, until an American astronomer named Gerald Hawkins revolutionized notions about the function of Stonehenge. Using a state-of-the-art (by 1960s' standards) IBM computer owned jointly by Harvard University and the Smithsonian Institution, Hawkins investigated the mathematical relationships between the holes and the stones at Stonehenge. His conclusion shook the worlds of astronomy and archaeology and helped crystalize a new discipline called archaeoastronomy.

Hawkins confirmed that not only is the location of the sun at the summer solstice predicted by Stonehenge's heelstone but the enigmatic station stones mark other crucial events in the astronomical calendar. These alignments correctly identify the most northerly points at which the sun and moon rise and the most southerly points at which the sun and moon set over the course of a year. Even more extraordinary, he showed that the fifty-six Aubrey Holes could be used to predict astronomical events such as the timing of the new moon's reappearance and the occurrence of eclipses. To predict the former, you simply move a marker two Aubrey Holes each day; when one circuit is completed, one full cycle of the moon, twenty-eight days, will also be completed.

Predicting eclipses using the Aubrey Holes and the station stones is a more complicated and long-term enterprise. It is perhaps the most extraordinary achievement by the builders of Stonehenge because it requires understanding the movement of two invisible points, the lunar nodes, in outer space. The lunar nodes are where the moon's orbit, which is slightly tilted with respect to the Earth's orbit around the sun, intersects the plane of the Earth's orbit. The four standing stones are nothing less than Neolithic calibration points for this 4 billion-year-old astronomical cycle. The fact that these nodes were recognized by a society that had not long emerged from the dark depths of a ten thousand-year ice age is nothing short of phenomenal.

Eclipses, whether solar (when the sun's disk is obscured by the moon) or lunar (when the moon's disk is obscured by the shadow of the Earth), can only occur when the sun is closely aligned with a node. Lunar nodes and eclipses are thus intimately linked. To keep track of the nodes, the designers of Stonehenge moved a marker by three Aubrey Holes per year (at every midsummer solstice). One circuit takes 18.67 years, exactly the amount of time for the moon's orbit around the Earth to intersect the ecliptic. The northeast orientation of Stonehenge and the mathematical symmetries of the Aubrey Holes show that this strange monument on an unremarkable grassy slope in southern England is a prehistoric calculator.

The most compelling evidence for Stonehenge's status as the world's first computer, however, comes from its physical location on Earth. In a relationship whose mathematical precision is almost eerie, the angle between the position of sunrise at the summer solstice and the most northerly moonset (marked by the most northerly of the standing stones) is a right angle, as is the angle between the position of the midwinter sunset and the most southerly moonrise (marked by the most southerly of the standing stones). Join the other two standing stones by a line, and that line's orientation toward the west predicts the position of sunset on the May and August quarter days—the four days in a year that herald the start of a new season—while to the east it predicts the position of sunrise on the February and November quarter days. There are only two regions in the world where this orientation can work, in two bands of latitude less than one-degree wide—one running through southern England between Oxford and Bournemouth and the other near the Falkland Islands in the South Atlantic. The chance that Stonehenge was accidentally located in one of only two such possible seventy-mile-wide bands on the face of the Earth is—literally—astronomically improbable.

Some ruffled feathers arose from Hawkins's discoveries about Stonehenge. For almost two centuries, the British archaeological establishment had been run as a type of private gentleman's club with a heavy emphasis on the arts. Hawkins's brash reclassification of Britain's most famous monument as a stone computer contrasted sharply with the British archaeological community's respectfully quaint designation of it as a Neolithic temple. There were two areas of contention: the first was that a cherished theory developed by generations of arts graduates had been tossed out, and the second was that the man responsible was a colonial. Neither was cricket!

The British archaeological establishment quickly closed ranks and drafted another astronomer to check Hawkins's calculations. Fred Hoyle, a man for whom the word idiosyncratic could have been concocted, was a maverick of the British astronomical establishment whose 1957 novel, *The Black Cloud* (about extraterrestrial invasion), had raised a few eyebrows and whose theories on the origins of life in the universe were regarded by some as verging on the crackpot. Fred

Hoyle, however, was British and, what was more, was nothing less than the Plumian professor of astronomy at Cambridge University. He was a safe pair of hands—or so the British archaeologists thought.

One evening, while dining at St. John's College, Cambridge, Hoyle's friend, the noted Stone Age archaeologist and mystery writer Glyn Daniel, mentioned Hawkins's book, *Stonehenge Decoded*. In a conversation that could only take place over the port and claret of an Oxford or Cambridge high table, Daniel asked Hoyle if he would check the astronomy in Hawkins's book. Daniel reasoned that, since Hawkins had made (at least in the opinion of the establishment) such a hash of the archaeology, might not his astronomy be at fault, too? Curious, Hoyle agreed to take a look at Hawkins's findings.

Hawkins's book had been written in a popular style, and Hoyle, who specialized in popular writing himself and had suffered the inevitable jibes from colleagues for it, readily understood why Hawkins had fallen afoul of the British archaeological establishment. Forearmed to see beyond Hawkins's rosy prose to the argument beneath it, Hoyle quickly confirmed Hawkins's arithmetic. After reviewing the work as a whole, Hoyle commented simply, "it seemed to me to have a genuine ring of truth about it."

Two parts of Hawkins's reasoning were the most compelling to Hoyle: the ability of the station stones to predict the rising and setting positions of the moon through its highly complicated 18.61-year cycle, and the fact that Stonehenge would be useless if it had been built only a few miles north or south of its critical latitude. Hoyle "found it impossible to dismiss this property as a mere coincidence."

Hoyle did take issue, however, with Hawkins's theory about the Aubrey Holes. He could not see how Stone Age man could have calibrated the system. Despite the fact that one circuit of the fifty-six holes almost precisely equals three 18.61-year-long lunar cycles, Hoyle correctly reasoned that one rotation of a marker around the circle would not always coincide with the occurrence of an eclipse at Stonehenge. Digging deeper into the mathematics, Hoyle soon discovered that careful observation of moonrises, coupled with the use of the Aubrey Holes, would predict all eclipses, although not all of them would have been visible at Stonehenge.

Hoyle's findings led him to conclude that the Aubrey Holes at Stonehenge comprised a Neolithic orrery that kept track of the positions of the sun and the moon in their endless orbital cycles as well as the major axis of the moon's orbit. Hoyle had gone native; the Plumian professor was a Stonehenge astroconvert. Far from placating British archaeology, Hoyle's agreement with Hawkins only further incensed the establishment. British archaeology, it seemed, suffered a catastrophic failure of imagination when faced with the prospect that Neolithic humans could comprehend math that the archaeologists themselves could not master. Hoyle, however, was a scientist and was not interested in the bleatings of numerically challenged arts graduates; he responded like the typically bluff Yorkshireman he was: "It is not speculation to assert that we ourselves could use Stonehenge to make eclipse predictions. We could certainly do so without making any substantive changes to the layout. While this does not prove that Stone Age men did in fact use Stonehenge for making eclipse predictions, the measure of coincidence otherwise implied would be quite fantastic."

Nicely put.

## THE FIRE OF HEARTH AND HOME

Let us turn to Stonehenge's power source, Sol, the star at the center of our solar system. The scale of our sun takes some understanding. It is over one hundred times the diameter of the Earth and has a surface area almost twelve thousand times that of our planet. The solar prominences that regularly arc from its surface could easily swallow the Earth.

Sol is a solitary yellow dwarf of spectral type G2. There are seven categories of stars (O, B, A, F, G, K, and M), which refer to their different positions on the rather fearsomely named Hertzsprung-Russell diagram. In 1913, Henry Norris Russell recognized the relationship between a star's temperature (routinely measured by assessing its color) and its brightness. To demonstrate the relationship, he designed a diagram with brightness on one axis and temperature on the other. Sadly for Russell, it was such a good idea that it was hit upon almost simultaneously by the German Ejnar Hertzsprung.

The Hertzsprung-Russell (HR) diagram is a fundamental statement about the way most stars make their living, that is, by the atomic fusion of hydrogen to helium within the nuclear furnaces of their superdense cores. It is a neat trick, clean and very efficient, and has been the holy grail of nuclear energy enthusiasts since the heady days of the 1950s when a hydrogen bomb—an inefficient relative of a star's interior—somehow became an interesting thing to test on the surface of the only habitable planet we know. The HR diagram demonstrates that stars are not randomly distributed in terms of brightness and temperature. They are confined to clearly defined regions on the diagram. The most important of these regions is the main sequence.

At the far lefthand side of the main sequence are the superhot, very bright, O-category stars. These stars usually have a surface temperature of 30,000 to 50,000 degrees C and are between twenty and one hundred times more massive than Sol. A good example is Alnitak, the left star in Orion's belt. At the other end of the spectrum are the M-category stars with surface temperatures of less than 3,600 degrees C and a mass less than half that of Sol. These are the red dwarfs, and we do not have to look farther than our nearest stellar neighbor, Proxima Centuri, to find an example. O-category stars are like future Nobel Prize winners: hot, rare, and bright. M stars are the football players: cool, numerous, and dim. In between these types lie all the other main sequence stars. G-type stars like our own Sol lie toward the dimmer, cooler part of the middle of the series. Not all stars lie on this main sequence. Off in left field, fizzing with high temperatures but dim luminosities, are the white dwarfs. At the top right are the giants and supergiants, with cool temperatures but bright luminosities.

One of the most interesting things about the HR diagram, although not obvious when it was developed, is that it is fundamentally about time. The stars on the main sequence evolve along its length from faint red protosuns to collapsing red giants that lead to red dwarfs. The various parts of the main sequence have been given names to reflect this stellar evolution. Our sun is currently about halfway along the main sequence in the so-called horizontal branch. When you next look up at Alnitak, recall that you are observing a sun still in the special-care baby unit, and when you contemplate Proxima Centuri, think about assisted living. Like every one of us, stars are born, grow old, and die. Stars, like all terrestrial and extraterrestrial bodies, have unique life histories.

The age of the solar system is broadly 4.5 billion years. In other words, this is the age of the sun, and the sun and the planets are the same age. But what was here before our solar system? The universe is 13 billion years old, so there was a long period before our solar system came into being.

More than 4.5 billion years ago, what would become the sun was part of a gas cloud with a diameter of more than fifty light years; compare this with the sun's current size—a little more than one hundred times the diameter of the Earth. It was not a dense cloud and contained only a few thousand atoms per cubic centimeter. Even so, there was enough material to make several solar systems. This vast gas cloud was cold, only a few degrees above absolute zero, and existed in a fragile equilibrium, neither expanding further into space nor contracting in upon itself. But something happened to disturb this fine balance. It seems that one of the stars in our local group finally gave up the interstellar ghost and went nova. After billions of years of tranquility, the pace of life in our proto-solar system began to pick up. Within only a few thousand years, concentrations of atoms began to form clusters within this giant cloud at the same time as its temperature began slowly rising. It was one of these clusters, randomly drifting in the cloud of dust, that would become our sun.

Within 400,000 years, a time span infinitesimally small against the scale of astronomical or geological time, the cluster had contracted to a core one millionth its original size, still making it over four times larger than the size of the solar system today. The center of the core was beginning to heat up as the dust particles were forced into closer and closer proximity. As that happened, the core began to radiate energy and, in so doing, began to slow the capture of more dust particles. The core was now a protosun. After only a few thousand years more, our protosun collapsed to something the size of Mars's orbit. The temperature had now reached 56,000 degrees C—the temperature at which atoms start to ionize—and the core began emitting a feeble red light. Our sun had arrived at the gates of the main sequence.

The contraction continued until the temperature was high enough to begin burning deuterium and forming helium–3, and the core ultimately reached the temperature at which hydrogen ignites. At that point, the protosun had become a star with a mighty fusion reactor burning fiercely at its center. During this stage, the sun was quite unstable, varying wildly in luminosity and surface temperature, but after 30 million years or so, its structure stabilized. The sun had started on its long journey across the horizontal branch of the HR diagram. At the present time, our sun is in its middle age, 4.5 billion years old and about halfway down the main sequence, with another 5 billion years to go.

What will happen then? When about 10 percent of the sun's hydrogen has been converted into helium, the nuclear fusion reactions that currently power it will cease. The electrons and nuclei in the core will begin to break down, and the sun will enter the asymptotic, giant branch of the HR diagram. The present equilibrium between the pressure directed outward from the burning core and the inwardly directed gravitation will fail, and the sun will start slowly collapsing under its own weight. The interior fires will reach outward to find fresh reserves of hydrogen in the outer shell, which will ignite. The outer layers of the sun will swell vigorously, swallowing Mercury and roasting Venus and Earth. Ultimately, the helium-rich core will be unable to lose heat fast enough, and it will become unstable. Within just a few hours, the core will overheat and explode. The outer layers of the sun will absorb the core's explosion, and the sun will now resemble a burning onion from which the center has been removed. The sun may repeat this cycle of contraction and expansion several times, but ultimately the carbon generated by these cycles will quench the core and prevent any further explosion. The outer layers of the sun, rich in helium, will continue to burn, and an intense solar wind will develop that will blow away the tattered fragments of the sun's outer envelope. After about 30 million years, the sun will expand one last time, swallowing Venus and Earth and losing as much as half of its mass to the interstellar darkness. The core will continue shrinking under the influence of gravitation, and all the remaining matter will collapse into a small body about the same size as the Earth but weighing fantastically more. The sun will become a white dwarf.

As it dies this white-dwarf death, our sun will draw from its tiny remaining reservoir of energy and emit a faint, red-tinged yellow light. Its atoms will be packed together so tightly that further collapse is impossible; a lump the size of a sugar cube will weigh as much as an automobile. It will continue to cool until it becomes as cold as the interstellar space around it. A good example of a white dwarf is the dim star Sirius B, companion to the much larger and brighter Sirius A. Having returned to the cold of its birth, ultimately the sun will emit no light at all. It will cycle endlessly through the

Milky Way as a carbon- and oxygen-rich black dwarf, which may eventually encounter another giant gas cloud and become involved in the birth of a new star, perpetuating the stellar cycle.

## THE STARMAN OF PADUA

Our sun is currently in middle age, with all the usual accoutrements of that comfortable time of life; it is well regarded at the center of the community with nine or, given the demotion of Pluto, now eight kids, sixty-odd grandchildren, and a stable income of hydrogen. Yet the sun leads a turbulent life. From time to time, its roiling surface is pocked with enigmatic black marks—sunspots, or dark clouds—that mar the otherwise equable face of the master of the solar system. Sunspots have been observed for almost as long as the sun itself. In China, records of sunspots predate the birth of Christ. Although Western observations are not as well documented, it seems likely that the Greek philosopher Anaxagoras discerned them as long ago as 467 BC.

In Aristotelian cosmology, sunspots caused a problem because they seemed to fly in the face of orthodoxy. The heavens were supposed to be perfect, and the notion of blemishes on the face of the sun was perilously close to heresy. It is no surprise, therefore, that only the biggest heretic of them all would initiate their scientific investigation.

Galileo Galilei was born on February 15, 1564, in Pisa, Italy, on the shores of the Ligurian Sea. Privately tutored at a young age, by 1581 he had matriculated as a student of arts at the University of Pisa, where his father wanted him to study medicine. His father's wish was a natural one—there was a distinguished physician in the family in the previous generation—but Galileo never took to medical studies, preferring mathematics and natural philosophy. In 1582 or 1583, Galileo encountered the mathematician Ostilio Ricci, who was to become a major influence on his life. Ricci successfully persuaded Galileo's father to allow him to focus on mathematics rather than medicine, and by 1585 Galileo was teaching at Siena. The following year, he traveled to Rome to visit Christopher Clavius, a German Jesuit who was a professor of mathematics at the Jesuit College and whose most notable claim to fame was that he assisted Pope Gregory XIII in introducing the Gregorian calendar. In 1589, Galileo, who was quickly building an enviable reputation as a mathematician, was appointed to the chair in mathematics at the University of Pisa.

In 1591, disaster struck the family. Galileo's father died, and the young scholar was faced with the prospect of supporting the family with his pitiful university salary. By 1592, he had landed a professorship of mathematics at the University of Padua for three times the salary he had received at Pisa. The subsequent eighteen-year period at Padua was later described by Galileo as the happiest of his life, and there is no doubt that it was his most scientifically fertile time. Galileo's main duties were to teach Euclid's geometry and the geocentric astronomy of the day. Ironically, his main audience was medical students who needed knowledge of the movements of heavenly bodies to assist them with astrology, a required part of their medical training.

At about this time, Galileo's heretical bent began to assert itself. In 1604, he gave three public lectures on the subject of a new star that had appeared in the sky, a supernova. By far the brightest in the sky and rivaling Venus in intensity, the new star was first noticed on October 9, 1604, and was observed by the astronomer Johannes Kepler on October 17. Remarkably, it was the second supernova to have been observed in a generation, for Tycho Brahe had observed one in 1572.

Like Brahe's supernova, Kepler's supernova (now known as Supernova 1604, after the year of its discovery) is located close by, within the constellation Ophiuchus in our own galaxy and only some 20,000 light years from Earth. Brahe's supernova, in the constellation Cassiopeia, is about twice the distance away, but both are our next-door neighbors in astronomical terms. Using parallax arguments, Galileo argued that Kepler's supernova could not be located anywhere within the solar system. The problem with this conclusion was that it implicitly criticized the geocentric tenets of Aristotelian astronomy. Galileo had become a closet Copernican, following Nicholas Copernicus, the Polish astronomer who pioneered the view that the Earth is in motion around its own axis as well as around the sun. For Galileo, it was the beginning of a long and slippery slope that would lead him to Rome and, ultimately, the Inquisition.

In May 1609, Galileo received a letter from his friend, Paolo Sarpi, describing an extraordinary new instrument recently exhibited in Venice and created by a Dutchman named Hans Lipperhey. Galileo later wrote:

> A report reached my ears that a certain Fleming had constructed a spyglass by means of which visible objects, though very distant from the eye of the observer, were distinctly seen as if nearby. [I applied] myself wholeheartedly to investigate means by which I might arrive at the invention of a similar instrument. This I did soon afterwards, my basis being the doctrine of refraction.

Although he did not invent the telescope, Galileo proceeded to build the one that would change history. By August 1609, after teaching himself how to grind and polish lenses, he had made an instrument vastly superior in quality to Lipperhey's, one capable of a staggering eight or nine times magnification.

Galileo was on the verge of discoveries that would revolutionize astronomy. In fact, during the eight weeks of December 1609 and January 1610, Galileo arguably made more discoveries that changed our world view than anyone has ever made before or since. So important were these findings that, in May 1610, Galileo published his first major work in astronomy, *Sidereus Nuncius* (Starry Messenger). The book caused a sensation with its tales of mountains on the moon, a galaxy (the Milky Way) made up of countless trillions of seedling stars, and four small, previously undetected bodies orbiting the giant planet Jupiter.

In part due to his alliance with the Medici family, in June 1610, Galileo became the chief mathematician at the University of Pisa and Signor Medici's personal mathematician and philosopher. At Pisa, Galileo became seriously interested in sunspots. He had already shown these dark regions on the sun's disk to a number of his friends during a visit to Rome in the spring of 1611. Not until 1612, however, did he initiate a systematic study of them. He was stimulated to do so by the activities of another Jesuit mathematician from the University of Ingolstadt named Christoph Scheiner. Scheiner began studying sunspots seriously in October 1611, and his first paper on the subject appeared in January 1612. Because Scheiner was a Jesuit, he was anxious to preserve the Aristotelian view of the perfection of the heavens; sunspots could not be blemishes on the face of the sun but instead must be orbiting bodies.

Scheiner was friendly with Marc Welser, a member of an old established family in the Augsburg region of southern Germany, who had a great interest in natural philosophy and who owned his own printing press. Welser published Scheiner's sunspot studies, and he sent a copy to Galileo in the winter of 1611. Soon afterward, Galileo suggested that sunspots were in fact marks on the surface of the sun or in its atmosphere rather than orbiting bodies. Although Scheiner agreed with many of Galileo's other ideas on astronomy, he would not accept that sunspots were not orbiting bodies.

In August 1612, Galileo formalized his ideas on the nature of sunspots by performing a large number of observations at the same time of day. In this way, the sun's orientation was standardized, and the motion of the spots across its disk could be easily discerned. Galileo's observations were facilitated by an invention of one of his research assistants, Benedetto Castelli. Castelli had devised a method of projecting the sun's image through a telescope, allowing its observation throughout the day and preventing the scourge of sun observers—eye damage.

Using Castelli's method, Galileo noticed that sunspots change their shapes and sizes and often seem to appear and then disappear on the solar disk, making it impossible for them to be orbiting bodies. The clincher, however, was the rate of movement of the spots across the sun's face. Galileo noticed that, although a spot took about two weeks to cross the sun, its rate of motion was not uniform. Specifically, the spot moved more slowly when it was near the edge of the sun than when it was near the center. Galileo correctly recognized this difference as a foreshortening effect. A spot nearing the edge of the sun is traveling either toward or away from you, and the resulting effect is of the spot moving more slowly across the disk of the sun. Conversely, when the spot is in the middle third of the sun, all its motion is lateral, and it appears to be moving quickly. This speed-related foreshortening effect can only occur if the spot is at or very close to the surface of the sun; a planet, by contrast, would not appear to change its speed while moving across the solar disk.

Galileo's three letters on sunspots were published in Rome in the summer of 1613, demolishing Scheiner's theories and his Aristotelian beliefs.

## THE LONE GUNMAN

The specter of global warming looms over us today exactly as the specter of nuclear holocaust loomed over everyone during the last century. Every day reveals more doom-laden scenarios as highly paid think-tank moguls and academics with an eye on their funding generate forecasts. Is it possible, though, that our ancestors, the same ancestors responsible for the exquisite positioning of the trilithons at Stonehenge, knew more about the importance of the sun than we do?

Galileo's sun observations and discovery of sunspots laid the foundation for a debate that still rages today and is relevant to our current concerns about climate change. To what extent is the sun responsible for Earth's climate changes? Why do periods of intense sunspot activity seem to correlate with warmer episodes on Earth, whereas periods of few sunspots correlate with cold periods?

Today we know that sunspots are areas of the sun's photosphere—its visible surface—that appear dark because they are cooler than their surroundings. Of course, "cooler" is a relative term when considering the surface temperature of a gigantic hydrogen fusion furnace. Sunspots are variously sized, but all are huge, anywhere between 2,500- and 250,000-km across and typically 1,500 to 2,500 degrees C cooler than the average surface temperature of the sun, a staggering 5,400 degrees C.

The sun has a layered structure that rivals the Earth's in complexity, although it lacks a definite surface. Its core is small (about 0.2 solar radii), dense, and very hot, with temperatures of up to 13.6 million degrees C. In contrast, the surface of the sun reaches temperatures of "only" about 6,000 degrees C. The core is where the main fusion reactions that produce the sun's energy take place, generating an energy equivalent to 900 million megatons of TNT per second. The rest of the sun is heated by this energy, which travels outward through successive layers from the core. The zone immediately outside the center of the sun, between 0.2 and 0.7 solar radii, is called the radiation zone, where the solar material is still too dense to allow heat transfers by convection. Instead, in this region, as the name suggests, heat transfer is via radiation. Further out still, from 0.7 solar radii to the visible surface of the sun, heat transfer is via convection. After these massive columns of hot gas cool, they sink toward the top of the radiation zone, where they pick up more energy and the cycle renews. The cycling thermal columns of the convection zone impart a characteristic granulated appearance to the sun's surface.

The visible surface of the sun is the photosphere, above which the sun's energy is free to escape into space. Immediately above the photosphere lies a narrow band, about 10,000 km thick, known as the chromosphere. Here the temperature varies widely, with a minimum of 4,000 degrees C about 500 km above the sun's surface, whereas in the transition region at the top of the chromosphere the temperature is almost 1 million degrees C. The name chromosphere means "color sphere," a name derived from observations made immediately before and after an eclipse when its baleful red light is particularly noticeable.

Beyond the transition zone is the corona, which gives off the spectacular streamers known as solar prominences. The corona, too, has a complicated structure. The K-corona (or continuum corona) interfaces with the chromosphere. This vast region, about the width of the sun itself, is created by sunlight scattering off electrons. Outside the K-corona is the E-corona (or emission corona), which contains abundant calcium and iron, beyond which is the F-corona (for Fraunhofer), created by sunlight bouncing off dust particles.

Sunspots are areas where strong magnetic fields emerge from the sun's surface. Sunspots usually occur in pairs of opposite magnetic polarity and are linked by loops of magnetic field that arc up through the solar chromosphere into the corona. The intense magnetic fields inhibit the convection of superhot material from the interior of the star, resulting in the lower temperatures that give them their distinct coloring.

The huge energy-creation processes within the sun generate strong electric fields that, in turn, create strong magnetic fields that extend far out into the solar system. Because the sun does not rotate at the same speed at the equator (about once every twenty-five days) as at the poles (about once every thirty-five days), the lines of the sun's magnetic fields become twisted together into a complex shape known as the Parker Spiral. Close to the sun, these contorted field lines cause sunspots where they penetrate the sun's surface, whereas away from the sun they bracket complex regions of space with differing magnetic polarity. The boundary between the differing magnetic polarities of the Parker Spiral in deep space forms a complex, rotating structure known as the heliospheric current sheet, which extends to the edges of the solar system.

Sunspots are complex beasts with an interior structure, a preferred habitat, and a disconcerting tendency to come and go in cycles. Larger sunspots have a central region known as the umbra, surrounded by a lighter halo known as the penumbra. In the umbra, lines of magnetic force tend to be vertically oriented, whereas in the penumbra, they are almost horizontal. Sunspots also tend to cluster within zones approximately thirty degrees north or south of the sun's equator. A sunspot's lifetime can be as short as a few days or, in the case of the larger spots, can persist for several months.

Many believe sunspots influence Earth's climate either directly or indirectly by signaling a pending change in solar intensity. The cyclicity of sunspots has excited curiosity ever since it was discovered by Heinrich Schwabe in a series of observations he made between 1826 and 1843. Schwabe was born in the German town of Dessau, where he trained first as an apothecary before turning his attention to astronomy. Curiously enough, what led Schwabe to discover sunspots was his quest for a planet that he believed was orbiting inside the orbit of Mercury. Reasoning that it would be extremely difficult to observe this planet, Schwabe decided to look for a dark spot passing across the face of the sun. For the next seventeen years on every clear day, Schwabe scanned the face of the sun watching for a spot moving across it regularly that was distinguishable from all the other blemishes cluttering its disk. All he saw were the spots that had been known since the time of Galileo. During this epic search, however, Schwabe noticed what appeared to be a regular variation in the number of sunspots. In an 1843 paper, Schwabe published his findings and suggested that the frequency of sunspots increased and decreased in a cycle of about a decade.

It was a revolutionary paper that excited the Swiss astronomer and mathematician Johann Rudolf Wolf, director of the Bern Observatory. Wolf continued Schwabe's observations and collated all the available data on sunspot activity from as far back as 1610, two years before Galileo's famous paper that started it all. Using this extended dataset, Wolf calculated the solar cycle as 11.1 years. In 1848 he devised a way of quantifying sunspot activity that is still used today. The Wolfer Number (known more generally today as the International Sunspot or Zurich Number) assigns a single value via a simple equation to the number and size of sunspots currently on the sun's disk.

It is not hard to see a relationship between sunspots and climate—the idea of a connection between the two is far from a new one. Indeed, in 1801 William Herschel (the Georgian astronomer who discovered Uranus in 1781) attempted to correlate the annual number of sunspots with the price of grain in London. During the northern hemispheric cooling known as the Little Ice Age (which lasted from 1450 to 1820), there were three periods of intense cold known as the Sporer Minimum (1450 to 1540), the Maunder Minimum (1645 to 1715), and the Dalton Minimum (1795 to 1820). Each of these periods correlated with times of few or no sunspots. The Dalton incidentally coincided with Herschel's observations and may have prompted his 1801 study of sunspots and grain prices.

By about 1900, it was recognized that there might be a relationship between the cyclicity of solar variations (of which the number of sunspots is the most obvious) and the Earth's climate. In America, Charles Greenley Abbot was assigned by the Smithsonian Astrophysical Observatory to investigate the relationship. So novel was the idea that Abbot had to start by inventing and building his own instruments to measure solar radiation. His efforts were worthwhile. As his research progressed, he discovered a number of cycles in the sun's solar radiation output, some of which were related to the 11.1-year Schwabe cycle.

As the twentieth century progressed, the Schwabe cycle was recognized as, in fact, only the tip of a hierarchy of cycles, all apparently based on the 11.1-year period. The twenty-two-year Hale sunspot cycle represents the two Schwabe cycles needed to return the sun's magnetic polarity to its starting point. But the Hale cycle also has its own peculiarities. For example, the polarity of positively and negatively charged regions on the sun's surface is the same in each hemisphere, but the northern regions are reversed with respect to the southern ones. Furthermore, the magnetic polarities of sunspot pairs reverse in each hemisphere between sunspot cycles, and as each cycle progresses, the sunspots converge on the equator.

Beyond the immediate Schwabe and Hale cycles are other, longer-term cycles. The eighty-seven-year Gleissberg cycle is approximately eight Schwabe cycles or four Hale cycles. At lower frequencies, there appears to be a 210-year Suess cycle and a 2,300-year Hallstatt cycle (although confirmation of these only came with the advent of much longer-term analyses of sunspot cycles in fossilized wood).

Studies of weather station data between 1861 and 1989 show a very high degree of correlation between the number of sunspots and the mean annual temperature of the northern hemisphere. Strangely enough, there is an even better correlation between the length of the sunspot cycle (which according to modern measurements averages 10.8 years but which quixotically varies between 9.5 and 11 years) and mean annual temperature. Shorter cycles correlate with lower temperatures, and longer ones, with higher temperatures.

All this information thus begs the important question of the exact relationship between sunspot cycles and Earth's climatic cycle and how it fits into theories of climate change. Clearly the transient darkening of the sun caused by sunspots is insufficient to reduce insolation and change Earth's climate directly. It seems that the increase in sunspot activity is associated with an increase in solar radiation hitting the Earth because the regions surrounding the sunspots become more energetic. How this translates into climate change on Earth, however, remains conjectural.

What is not conjectural is that variations in solar output do affect the Earth, most especially in the stratosphere, part of the upper atmosphere. Times of high sunspot activity are highly correlated with times of high solar magnetic activity, which deflects cosmic rays from hitting the outer atmosphere. According to Henrik Svensmark, a Danish climatologist, fewer cosmic rays lead to a decrease in condensation nucleii in the atmosphere and, hence, a decrease in cloud cover and a corresponding increase in Earth's surface temperature, for less sunlight is reflected away. The converse is also true. At the present time, the data suggest that sunspot activity warms the Earth's surface to about the same extent as greenhouse gas emissions and *on about the same timescale.*

A skeptic would say that such a statement is only as good as the data it is based on, and a well-informed skeptic would point out that the data only go back about as far as the last 150 years. This interval is far too short to make reliable estimates of the sun's impact on geologically significant timescales—the only timescale that matters when considering questions of this nature.

Thankfully, we can augment the timescales of human observation by mining the rich terrains of archaeology and paleontology. In 2004, an important study was published in the prestigious journal *Nature* that compiled data on the intensity of sunspot activity for the last 11,000 years (recall that Wolf had only managed to study as far back as the seventeenth century). The compilation was assembled using a variety of proxies for sunspot activity: specifically, the concentration of the isotopes carbon–14 and beryllium–10 in tree rings.

The study of tree rings, dendrochronology, is an old but very elegant way of ordering the sequence of past environmental events. The number and width of tree rings are highly dependent on the environment in which the tree grew. During wet summers, trees experience greater growth, and correspondingly, tree rings are wider than during dry summers. The consequence of this dependence on the environment is that each tree or group of trees builds up a unique signature of tree ring morphology—the number and width of rings—that can be categorically identified. It is comparatively simple to join overlapping tree ring records from similar trees—say, the same species from the same forest—and assemble a continuous record of climate change far into the past. Using long-lived trees such as the sequoia, this technique can extend the record into the relatively remote past (several thousand years).

Dendroclimatology has been conducted in this manner for decades, but the advent of radiocarbon dating has made the system even more accurate by providing tie-points in time. The concentration of carbon–14 in the atmosphere correlates to sunspot activity. During times of low sunspot activity, more cosmic rays hit the Earth's atmosphere because the solar wind is less and does not deflect them as much. This heightened cosmic ray flux correspondingly increases the production of cosmogenic isotopes (that is, isotopes such as carbon–14 and beryllium–10 that are formed by cosmic ray activity). Thus, during times of low sunspot activity, more carbon–14, for example, finds its way into living tissues and shows up in tree rings. The converse is also true: during times of high sunspot activity, the amount of carbon–14 incorporated into tree rings decreases.

The authors of the *Nature* study calibrated their scheme by comparing it with the known record of sunspot activity dating back to the seventeenth century and found a remarkably good match. On this basis, their eleven thousand-year record is likely to be right on the money, too. Through their research, the authors discovered that the past seventy years (until the present day) has been a time of anomalously high sunspot activity. Using Svensmark's model, this might account for today's higher temperatures, because high sunspot activity means fewer cosmic rays hitting the stratosphere, which in turn means fewer clouds and less reflection of sunlight. As a result, more solar energy hits the Earth's surface, thereby warming it up.

Where does this leave the current debate about greenhouse gas emissions? It perhaps shows that human-produced greenhouse gases are only one side of the global warming equation and that, without considering the effect of sunspots, we will be incapable of understanding our climate. Although the jury is still out on these issues, there is no doubt that Earth's climate is much more complex and influenced by much more than has been previously conceived.

The sun has been the subject of many space missions since the beginning of the robotic age of space exploration in the late 1950s. The earliest probes were of the Pioneer series—discussed in detail later (especially in regard to Venus and the outer planets)—which made the first detailed measurements of the solar wind and magnetic field. In the 1970s, Germany and the United States collaborated on the two *Helios* probes that passed inside the orbit of Mercury and set the record for the closest approach to the sun. The NASA *Skylab* project, which used the surplus equipment left over from the Apollo program, provided scientists with new data about the sun, particularly on the structure of the transition region and the corona. In 1980, NASA launched its *Solar Maximum* mission to investigate solar flares and showed that, contrary to expectations, the sun is actually brighter during the sunspot cycle maximum.

Japan, a relatively new but important player in the exploration of the planets, launched its own mission to the sun in 1991 with its *Yohkoh* (Sunbeam) satellite. It was the first mission to observe solar flares at X-ray wavelengths and allowed scientists to describe several different types of flares and to show that the sun's corona is far more active than had previously been thought.

The *Solar and Heliospheric Observatory* (*SOHO*) was a joint NASA/European Space Agency (ESA) project launched on December 2, 1995, which took up position at the Lagrange point between the

Earth and the sun some months later (a Lagrange point is a spot in space where the pull of gravitation between two objects is equal). What was originally planned as a two-year mission has now operated successfully for over ten, and *SOHO* has become a major source of space weather data related to changes in the sun's activity.

The joint ESA/NASA *Ulysses* probe was launched in 1990 to study the sun's poles. It showed that the sun's southern magnetic pole is much more dynamic than had been predicted and that its location is not fixed.

The *Genesis* mission, launched in August 2001, was the first to attempt to recover a sample of solar wind, the stream of charged particles ejected from the sun. For three years the spacecraft's collector arrays were extended to capture particles of the solar wind and were then stowed in a return capsule for the journey back to Earth. A drogue parachute failed to open properly, however, and the spacecraft crashed into the Dugway Proving Ground in Utah at almost two hundred miles per hour. Amazingly, some of the collector arrays were apparently undamaged and are currently undergoing investigation. It may be possible to salvage the mission's science after all.

Our knowledge and understanding of the sun has come a long way since our remote ancestors dragged colossal stones from Wales to Stonehenge in their attempt to understand local astronomical and solar cycles. Today, the new science of archaeoastronomy strives to interpret and extend Hawkins's and Hoyle's Stonehenge findings. It now seems certain that stones at Carnac in Brittany are even older (ca. 4,500 BC) than those at Stonehenge, and there are other important megalith sites at Newgrange in Ireland, Pentre Ifan in Wales, and Fossa in Italy. The Big Horn Medicine Wheel near Sheridan, Wyoming, was constructed fifteen hundred years before the birth of Christ, yet the spokes of the wheel are oriented to pinpoint the date of the summer solstice as well as the positions of the stars Aldebaran, Rigel, and Sirius. The great pyramid at Giza in Egypt, as well as the Inca temple at Machu Picchu, seem to have astronomical significance as well. To these ancient civilizations, the sun was the centerpiece of the universe, and as we face the challenges of understanding the Earth's future, we should not forget that wisdom.

# 2

# The Piper at the Gates of Dawn: Mercury

In the early 1960s as the space age got underway, America and the Soviet Union were looking for planets to explore with their new missile and space technologies. The reason was partly political; after the Bay of Pigs and the Cuban Missile Crisis, the Cold War was at freezedown, and both the Soviet Union and the United States viewed the conquest of space as a metaphor for the political struggle between communism and capitalism here on Earth. The most obvious manifestation of the struggle was the manned space programs, most memorably the Apollo program, but the unmanned programs, the robotic missions, were getting underway as well. The real scientific value of robotic missions would only be appreciated after many years, but in the meantime, they, too, played a part in the grand political game.

The obvious candidates for the earliest robotic missions were the planets that bracketed Earth: Mars and Venus. But quite early, in 1963, scientists realized that an extraordinary launch opportunity would open within a decade to the planet closest to the sun—Mercury.

In those days very little was known about Mercury. The scientific thinking at the time suggested that Mercury's proximity to the sun would make it the mineral treasure house of the solar system. It was an idea that permeated the science fiction of the time, too, as in Isaac Asimov's classic short story "Runaround." As we shall see, this thinking persists to this day.

## THE WINGED MESSENGER

Runaround is a good name for Mercury, for it orbits the sun in only eighty-eight days and, hence, appears to flit very quickly in and out of the sun's glare. The Greeks gave it two names, Apollo and Hermes, because they thought it to be two separate planets, one visible at sunrise and the other at sunset. In Greek mythology, Hermes was the fleet-footed messenger of the gods, a name that reflects Mercury's rapid movement across the sky. The Romans named the planet after their god of trade, profit, and commerce—their counterpart for Hermes—and this is the name we still use today.

Mercury was known even before the Greeks and Romans by the Germanic peoples and Scandinavians who navigated far from their native lands into the waters of the Mediterranean and the northern coast of Africa. During the balmy dawns and dusks of the preclassical world, they watched the small planet loop upward from the horizon and then swiftly back down, and they knew they were seeing something very strange, indeed. Their names for Mercury were Wodan and Odin.

The Mesopotamians, who lived in the triangle of land between the Tigris and Euphrates rivers (present-day Iraq), also observed Mercury. They knew that the interval between successive reappearances of Mercury in the same area of sky was shorter than for the other planets and correctly reasoned that it moved more rapidly than the others. They named it Nebo, after the Mesopotamian god of wisdom and writing who kept the book of fate. The Greek astronomer Eudoxus was the person who, in about 400 BC, measured the period between successive appearances of Mercury in the morning versus the evening sky as 115 days and realized, correctly, that it was a single planet.

After the invention of the telescope in the seventeenth century, Mercury proved an elusive target for the new breed of "optick tube" astronomers. Because it is never more than twenty-eight degrees from the sun (as viewed from Earth), it has to be observed either at twilight or during the day; it cannot be seen at night. Observation at twilight carries the advantage of a good contrast between the disk of the planet and the relatively dark background of the sky, but its disadvantage is that the path length between the telescope and the planet (through a hazy and turbulent atmosphere) is relatively great, degrading the view. Daytime observations allow for a shorter path length through the atmosphere, but the lower contrast between the planet's disk and the surrounding sky make Mercury hard to find. Even with today's instruments, the best viewing of Mercury is still in the early morning or late evening when it is not obscured by its gigantic, hydrogen-fusing neighbor.

In addition to these difficulties, there are only about thirty or forty days in the year when Mercury's angular distance from the sun allows it to be viewed by telescope from the Earth. Coupled with the fact that Mercury is an inferior planet (that is, closer to the sun than the Earth) and therefore goes through phases like the moon and Venus, it is a wonder that, even by the 1960s when the Mercury launch opportunity was discovered, we had much information at all about the planet.

The two greatest observers of Mercury were Giovanni Schiaparelli and Eugene Antoniadi. Schiaparelli pioneered the technique of observing Mercury during the day. Beginning in 1881 and avoiding tobacco, alcohol, and coffee lest they dull his senses, Schiaparelli pointed his telescope at Mercury, intending to take the first steps in transforming a planet into a place—by mapping it. The task was heartbreaking and frustrating, and after eight years of observation, he could report only "extremely faint streaks" on the planet's surface. He sketched these into a rough map of Mercury that he published in 1889. One aspect Schiaparelli did notice, however, was that the visible features of Mercury never seemed to change. This observation started an idea that would hold sway for most of the twentieth century—that Mercury is gravitationally locked by its proximity to the sun, keeping one face perpetually toward Earth and the other toward the sun. Mercury's popular depiction, therefore, was as the foundry of the solar system, with one face eternally broiling, the other eternally frozen, and an intermediary twilight zone between the two.

Eugene Antoniadi was arguably the most influential astronomer of the early twentieth century. Born in Turkey but of Greek descent, he spent most of his life in France and was the lynchpin of French astronomy between 1900 and 1930. He is most famous for his role in debunking Schiaparelli and Lovell's canal theory of Mars, but he also furthered Schiaparelli's famous Mercury map by using the massive telescope at the Meudon observatory outside Paris. In 1934 Antoniadi released his own version of the map that, although including little new information, showed amazing and largely imaginary detail of the planet's surface. The faint streaks seen by Schiaparelli were resolved by Antoniadi into geographic features that Antoniadi was unable to resist naming: Caduceata and Apollonia were topographic lows ("seas") that dominated the southern hemisphere, whereas the "highlands" of the northern hemisphere he named Solitudo Hermae Trismegisti, Premethei, and Persephunes.

Like Schiaparelli before him, Antoniadi was convinced that Mercury showed only one face to the sun. Not until after the dawn of the space age was this myth finally laid to rest. The notion had profound implications, however, for Schiaparelli and Antoniadi reasoned that life could exist in the twilight zone between the two faces of Mercury, especially if, as Schiaparelli believed, the planet exhibited an extreme libration (wobble) that maximized the zone of eternal dusk. But Schiaparelli and Antoniadi were wrong, for Mercury does not show one face eternally to the sun and the other to an indifferent solar system. They were fooled by one of the strangest rotational regimes in the solar system combined with the infrequency with which Mercury can be observed from Earth.

## TRAVELS IN THE TWILIGHT ZONE

Schiaparelli's suggestion that Mercury's day is the same length as its year (a 1:1 resonance) dominated the scientific and popular consciousness about the planet for almost a century. Schiaparelli's model of Mercury began to come unglued in 1962, however, when a Michigan-based astronomer named W. E. Howard and his colleagues showed that the night side of Mercury—the one supposedly eternally frozen—is warmer than it should be if it never receives any insolation from the sun. In 1965, these observations were refined by Gordon Pettengill and Rolf Dyce of Cornell University using measurements from the Puerto Rico–based Arecibo radio telescope, the largest single-dish radio telescope in the world. Pettengill and Dyce showed, beyond doubt, that Mercury is not gravitationally locked into a 1:1 relationship with the sun. Instead, its day is about 58.6 Earth days, or two-thirds the time it takes the planet to orbit the sun, giving an orbital resonance of 3:2. Because of the limited number of opportunities for observing Mercury, this spin-orbit resonance has the curious effect of favoring one view to Earth-bound observers, thus leading to the erroneous supposition of a 1:1 orbital resonance.

Other bodies in the solar system do exhibit 1:1 gravitational coupling, however. We always see the same side of the moon because of this phenomenon. The effect is thought to be caused by the gravitational effect of the primary body, the object with larger mass, in this case, the Earth. The inner five moons of Jupiter also show a 1:1 spin-orbit resonance. However, as the case of Mercury shows clearly, 1:1 resonances are not the only possible gravitational locks.

## THE WANDERER AT THE EDGE OF TIME

NASA was formed on July 29, 1958, as a direct response to the crisis engendered by the Soviet's perceived superiority in space technology, evidenced by the launch of *Sputnik 1* on October 4, 1957. Within only a few years, NASA's primary objective became to land a man on the moon by the end of the 1960s, a commitment made by President John F. Kennedy on May 25, 1961. Consequently, the earliest missions sent out by the Jet Propulsion Laboratory, NASA's prime unmanned spacecraft contractor (and, as we shall see, soon to become a major thorn in the side of NASA), were the lunar *Ranger* probes, designed to take close-up photographs of the lunar surface and send them back before crashing into the moon and self-destructing.

The early Ranger missions were notably unsuccessful and were conducted in the full glare of international publicity as the United States tried to catch up with the Soviets in the space race. For a long time, the Soviet Union dictated the terms of that race. In early February 1961, they dispatched *Venera 1*, the first in a long series of very successful probes to Venus. The Jet Propulsion Laboratory responded by modifying the basic design of the Ranger series to come up with a probe for deep-space reconnaissance—the *Mariner* probes.

The Mariners are a large family of space probes that we shall meet many times in the course of this book, and to get acquainted, it is worth quickly summarizing their long and colorful history. *Mariner 1* was launched from Cape Canaveral Air Force Base on July 22, 1962, but malfunctioned and had to be destroyed by the Cape's range safety officer. *Mariner 2* was launched on August 27, 1962, and after a 3½-month cruise became the first spacecraft to successfully fly by another planet; it sped past Venus on December 14, 1962, confirming that planet's backward spin, high surface temperature, and thick, carbon dioxide–dominated atmosphere. *Mariners 3* and *4* were aimed at Mars. *Mariner 3* failed when the vehicle's nose fairing did not jettison after launch, whereas *Mariner 4*, as we shall see, became the first spacecraft to fly past the red planet—albeit with dispiriting results.

*Mariner 5*, originally a backup for *Mariner 4*, was re-tooled and sent instead to Venus for a close-up look, arriving there in October 1967. *Mariners 6* and *7* consisted of the first dual mission to Mars, and both were very successful, flying past the planet at an altitude of only slightly more than two thousand miles, analyzing the composition of the Martian atmosphere and sending back hundreds of pictures. *Mariners 8* and *9* were identical and intended to orbit Mars and map its surface in great detail as a prelude to the Viking lander missions scheduled for 1975. Although *Mariner 8* failed, *Mariner 9* successfully entered orbit in May 1971 and became the red planet's first artificial satellite. *Mariner 9*'s images revealed massive volcanoes such as Olympus Mons (the largest volcano in the solar system), channels possibly carved by running water at some point in the planet's past, and a massive canyon system, including a vast gash 3,000 miles long, 120 miles wide, and up to 4 miles deep. The latter extraordinary feature—one of the defining characteristics of Mars—was named Valles Marineris in honor of *Mariner 9*'s extraordinary achievement.

*Mariner 10,* the last of the Mariners, was a departure from the rest of the series, for it was sent to Mercury. It was an extraordinarily innovative mission in that it was the first to use the technique known as "gravity assist." The idea is for a spacecraft to pick up energy by parasitizing the gravity-well of conveniently located planets en route. In other words, if a spacecraft falls toward a planet, it will pick up speed. At the correctly calculated moment, igniting the spacecraft's engines will boost it out of orbit, maintaining the added energy imparted by its passage through the host's gravitational field. The Jet Propulsion

Laboratory and NASA would use this technique widely throughout the succeeding decades.

For *Mariner 10,* the gravity assist was to be provided by one of the most enigmatic planets of the inner solar system: Venus. At the time of the *Mariner 10* mission, Venus was perfectly positioned for the Mercury-bound spacecraft to slingshot around. The Venus gravity assist slingshot had two beneficial consequences: a much smaller (that is, cheaper) launch vehicle would suffice, and it would be possible to make at least one (and feasibly more) return visits to Mercury after the initial flyby. The slingshot technique was a way of getting the maximum value for the money out of the mission.

The *Mariner 10* was by far the most complex mission attempted to that point. The launch window was extremely narrow, between October 16 and November 21, 1973. NASA elected for a November 3 launch because then *Mariner 10*'s first encounter with Mercury would be at quadrature, when the planet would be half-lit by a sun low on its horizon, maximizing contrast. The long shadows cast by the sun would help distinguish the detail of Mercury's surface.

*Mariner 10* was an "all-up" mission, fully loaded and with no expense spared—the first of NASA's "golden-age" planetary exploration missions. It carried a suite of seven scientific experiments. The first and most important of these involved television imaging. After the frustrations of a century of inadequate Earth-based telescopic observation of Mercury, and especially after the orbital resonance fiasco, scientists were in no mood to let this wonderful opportunity go to waste; *Mariner 10* was equipped with two state-of-the art vidicon cameras for photographing the geography of Mercury's surface in great detail. The features half-imagined by Schiaparelli and Antoniadi would at last be revealed.

The spacecraft also carried an infrared radiometer to measure thermal emissions from Mercury, two ultraviolet spectrometers to test the atmospheric composition, two magnetometers to test for a magnetic field, and two plasma detectors to measure the intensity of the solar wind and to observe the effect of high-energy particles on the planet. Finally, the spacecraft's elaborate radiotelemetry system, which enabled the spacecraft to communicate with Earth, would also measure Mercury's mass and diameter using the measured gravitational deflection on the spacecraft as it flew by the planet. From these data, scientists would make inferences about the planet's composition and internal structure.

The pre-launch was tense for the *Mariner 10* scientists and engineers, because NASA had decreed that there would be no duplicate spacecraft launched. At this time in the history of unmanned space exploration, given the many failures that attended the missions to Mars, Venus, and the moon in the 1960s, NASA commonly launched a backup spacecraft in the event of a failure with the first. Backup spacecraft were relatively cheap to produce, and they saved a lot of angst when things went wrong. But it had been decreed that only one spacecraft would be dispatched to Mercury, with the standby left on the ground at Cape Canaveral in Florida. The spare would only be launched in the event of a catastrophic mishap to the first. With a launch window as tight as *Mariner 10*'s, everyone knew there was a real risk the second spacecraft/launch vehicle could not be readied in time and the mission would be lost.

*Mariner 10* was launched atop an Atlas/Centaur rocket from Launch Complex 34 at Cape Canaveral on November 3, 1973. After launch, *Mariner 10* briefly paused in orbit while the science instruments were calibrated. Then the spacecraft fired its hydrazine engine and headed for its gravity assist encounter with Venus. As the program managers breathed sighs of relief and the backup launch window closed, almost immediately mission engineers began to experience problems with the spacecraft. As *Mariner 10* headed toward Venus, she began to experience serious problems with her flight data systems, and the engineers feared a problem with the spacecraft's power system. In addition, part of the high-gain antenna system failed, and it seemed as though picture quality from Mercury would be severely compromised.

Despite these and other problems, *Mariner 10* successfully reached Venus on February 5, 1974. She took more than four thousand pictures of Venus's atmosphere before passing out of range, continuing to her primary target, which she encountered forty-three days later.

En route, the spacecraft's approach toward the sun started to cause surface heating, and much to the consternation of her designers, she began shedding particles into space. This type of problem had not been encountered before, but then no other spacecraft had ever functioned so close to a star. One of the particles vaporizing from *Mariner 10*'s aluminum skin drifted in front of the star-tracking apparatus needed to keep the spacecraft orientated. The craft lost its lock on its celestial reference point, the star Canopus. It took almost two hours for the spacecraft to reacquire the star, during which time the spacecraft's gyros were actively swigging precious maneuvering fuel. As more and more particles drifted in front of the star-tracker, the problem repeated, and mission engineers worried about the spacecraft's fuel reserves.

In a spurt of creative frenzy reminiscent of the great days of the Apollo moon landings, when engineers seemed to possess an almost supernatural empathy with their spacecraft, those in charge of *Mariner 10* found an ingenious way to steer the craft that required no fuel at all—or none that the spacecraft had to supply itself. They used the energy of the sun. By tilting the solar panels, they found they could vary the pressure of the solar wind—the stream of photons flowing from the sun—on the spacecraft and thereby control its trajectory through the interplanetary void. The *Mariner 10* engineers named the technique "solar sailing." It was a singular triumph and one that, in later years, would become of major importance in flying spacecraft when fuel constraints were critical.

Meanwhile, the trajectory alignment at Venus had been so accurate that, as the spacecraft neared Mercury, only a single course correction was needed to swing *Mariner 10* over to Mercury's night side and position her correctly for her return visit to the day side in six months. Because of impoverished maneuvering fuel, the course correction was performed with a single, critical main engine burn to nudge the spacecraft into the desired orbit. On March 16, 1974, the maneuver was successfully completed, and the next day the science experiments were turned on in preparation for *Mariner 10*'s historic observations of Mercury.

## THE HERMIAN DYNAMO

At its closest approach, *Mariner 10* was only seven hundred kilometers above Mercury's surface. The craft's first task was to search for an atmosphere by observing ultraviolet solar radiation and thermal infrared omission from the ground. It found that Mercury has virtually no atmosphere—a few wisps of noble gases are all that is left of any atmosphere the planet might once have had. But *Mariner 10* did find something no one expected: a magnetic field.

Not all planets have a magnetic field, and not all magnetic fields are the same. Earth's magnetic field is particularly complex, with a clearly defined north and south magnetic pole (that is, the field is dipole). The field performs a crucial function by protecting the Earth and the tenuous skin of life clinging to its surface from the ravages of the sun's high-energy particles. The Earth's magnetic field is thought to be a consequence of its dense liquid metal core, which, by spinning, acts in much the same way as a dynamo. Scientists believe that, for *any* planet to have a magnetic field, two conditions must be satisfied: It must have a liquid core, and it must rotate rapidly. The faster a planet rotates, the faster the liquid core rotates, and the stronger the magnetic field produced by the moving charged particles within it. But a planet's likelihood of having a molten core is related to its size.

Small planets (those with a high surface area to volume ratio) generally have radiated their interior heat out into space long ago. Large planets are more likely to have retained it. Earth maintains a magnetic field because it is a relatively large planet with a rapid rotation. Venus, although about the same size as Earth, rotates very slowly (once every 243 Earth days), and hence it does not have a magnetic field. Mars spins quickly (in a little under twenty-five hours) but is very small and so has little interior heat; consequently, it has only a weak magnetic field. But Mercury presents a paradox. It is very small and rotates slowly, so we would not expect it to have a magnetic field. The fact that it does strongly suggests that it has a molten core, thereby demonstrating an unexpected kinship with our own planet. Mercury's magnetic field is not as complex as our state-of-the-art dipole field, but, nonetheless, it is there.

Mercury is now known to have by far the largest core relative to its size of any planet in the solar system. Its mass is 60 percent greater than can be predicted by its size alone. This mass probably results because Mercury's interior is awash with molten heavy metals—much like the Earth. Perhaps when the time comes for human interstellar travel, we will head to Mercury, the piper at the gates of dawn, for the metals necessary to build our starships.

## ENCOUNTER AT NEARPOINT

After discovering the Hermian magnetic field, *Mariner 10* flew around the sun and prepared for its second encounter with Mercury, this time flying across the planet's southern hemisphere. By now the spacecraft's altitude was 50,000 kilometers, a distance that allowed it to fill in the imaging gaps from its first observations.

The third and final flyby was the closest by far and, like the first, was across the dark side of the planet. The spacecraft skimmed only 330 kilometers above the surface, with instructions to search for the source of the magnetic field it had found on its first flyby. The extremely low altitude allowed the craft to capture the highest resolution pictures of the entire mission. At times, near the darkside/lightside border, the craft could resolve objects a mere ninety meters apart.

*Mariner 10* continued to collect data until March 24, 1975, when its already impoverished supply of maneuvering gas was finally exhausted. The mission, the only one to Mercury to date, lasted 506 days. *Mariner 10* is still in orbit around the sun, even though its ancient electronic systems have long since been cooked by deadly solar radiation. *Mariner 10* was an acid test not only of the materials of U.S. spacecraft construction but also of the management skills of the engineers and scientists at the Jet Propulsion Laboratory. The high radiation environment, coupled with the complexities of the first-ever gravity assist maneuver, could easily have derailed the mission. By any standards, the *Mariner 10* mission was an incredible success, especially coming so early in the U.S. space program.

## GROUND ZERO

The pictures returned by *Mariner 10* ranged in resolution from several kilometers down to a few tens of meters and showed a terrain superficially similar to that of the moon, covered with craters and lava plains. They suggested to the *Mariner 10* imaging team that Mercury had been subjected to heavy bombardment early in its history, followed by an interval of intense volcanism. Mercury's craters range in diameter from only a few meters to hundreds of kilometers; the largest is the giant Caloris Basin.

The Caloris Basin is probably the result of an ancient impact that occurred during the Heavy Bombardment Period of the early solar system, some 4.5 to 3.8 billion years ago. At a staggering width of 1,350 kilometers—840 miles—it is one of the largest impact craters in the solar system; the diameter of the impacting body must have been at least one hundred kilometers. The Caloris Basin is similar to the giant Imbrium Basin on the moon, which is 1,100 kilometers wide. The name Caloris Basin means "basin of heat" and refers to the fact that the temperature within the basin can reach 427 degrees C when the planet is at its closest approach to the sun. So severe was the impact that the concentric ridges surrounding the Caloris Basin are almost two kilometers high, although the floor of the basin itself is covered with smooth plains known as the Caloris Planitia. Precisely at the antipode of the Caloris Basin, that is, at the exact opposite side of Mercury, is an unusual, tormented region of hilly terrain known as the "weird" or "chaotic" terrain. Scientists have ventured two suggestions for the formation of this strange topography, and both are related to the formation of the Caloris Basin. One suggests that the shock waves from the colossal impact focused either *through* the planet or *around* its circumference and exited or met on the opposite side of Mercury, fracturing the surface. The other suggestion is that the terrain formed when the ejecta from the impact converged upon itself after circumnavigating the entire globe of Mercury.

The chaotic terrain is not the only unique feature on Mercury's surface. For example, Mercury's plains fall into two distinct categories—those with relatively few craters, and those with a lot. The plains may therefore have been formed during two distinct periods, with the less heavily cratered plains—the younger ones—having been formed by lava outpourings that buried older, more heavily cratered plains. Then there are great curved cliffs several kilometers high, known as lobate scarps. These lobate scarps or rupes (*rupes* is Latin for cliff) are interpreted as wrinkles on the skin of Mercury that formed as the planet's core cooled and the surface contracted in the millennia after its formation. The largest known rupes is Discovery Rupes, 550 kilometers (350 miles) long and 1.5 kilometers (about a mile) high.

That these surface features are still visible after billions of years is a clue to a major difference between Mercury and the other terrestrial planets: Mercury does not have and probably never did have a mechanism for renewing or modifying its surface. Earth still has active plate tectonics, Mars probably had tectonics in its distant past, and Venus, as we shall see, is dominated by more-or-less continuous volcanism. Mercury has neither of these processes, and as a consequence, its features are eternally etched on its surface.

## THE AGENT OF SPACE-TIME

Despite its strange, broken, and tortured terrain and its worrying proximity to a star—although the one we love and know best—Mercury, in a surreal way, provided the proof for one of the most dramatic theories of twentieth-century physics—Einstein's general theory of relativity. The roots of this story, one of the strangest in planetary science, go back to the middle years of the nineteenth century when Urbain Leverrier, a French mathematician, became interested in orbital perturbations. These small irregularities in the planets' movements around the sun are usually caused by the presence of other bodies. An obvious example is the geometry of the Earth/moon system. The moon's orbit is influenced by the mass of the Earth, and the Earth, although much larger, is reciprocally influenced by the mass of the moon.

Leverrier, who started studying planetary science in 1838, revolutionized our understanding of the way objects in the solar system react to each other. His work led him to believe that he could predict the existence of hitherto undiscovered planets on the basis of mathematics alone. In 1845, for example, he learned from the director of the Paris observatory, Francois Arago, of certain observed anomalies in the orbit of the newly discovered planet Uranus. Arago encouraged Leverrier to investigate the anomalies. Through calculus-based investigations, Leverrier predicted that there was another planet beyond Uranus. This planet, which came to be called Neptune, was independently observed by the Englishman John Adams and the German Johann Galle a few months later (although, as we shall see, the priority of the discoveries was by no means straightforward).

Encouraged by the success of his planet-locating calculus technique, Leverrier turned his attention to Mercury, which exhibited its own strange orbital perturbation. Mercury is unusual among the planets of the solar system in that it has an extremely elongated, or eccentric, orbit. In the language of celestial mechanics, its aphelion (the point in its orbit furthest from the sun) and its perihelion (the point in its orbit closest to the sun) are very dissimilar. The point in space occupied by Mercury's perihelion is not fixed, however. Mercury's perihelion moves, or precesses, as a result of gravitational interactions with other bodies.

Imagine the orbit of a planet as an ellipse (which most are), with Mercury's as a particularly elongated ellipse. If the sun occupies a space not perfectly in the middle of this ellipse, the two ends of the ellipse are at different distances from the sun; the aphelion is at the end of the ellipse furthest from the sun, and the perihelion is at the end of the ellipse closest to the sun. Draw a straight line between the two ends of the ellipse, and the movement of the perihelion will cause this line itself to rotate about the sun. This movement is called precession of the perihelion.

Mercury is not the only body in the solar system whose perihelion precesses. For any celestial body, the amount of precession can be calculated accurately using Newton's equations of motion—except for Mercury's. Mercury's perihelion advances by 476 arc-seconds per century, *which is 43 arc-seconds per century more than predicted by classical Newtonian mechanics.* This discrepancy was a major conundrum for nineteenth-century astronomy, and Leverrier was convinced it was caused by the presence of another planet lying closer to the sun than Mercury and perturbing its orbit.

For decades, astronomers in the late nineteenth and early twentieth centuries labored to locate this mythical planet, which Leverrier named Vulcan for the god of fire and forge. The search came to an abrupt end in 1915 when Einstein recognized that gravity is a form of acceleration and that mass and the space containing that mass have a reciprocal effect on each other. The physicist John Wheeler summarized the relationship succinctly: "Space tells matter how to move; matter tells space how to curve."

Einstein predicted that when Mercury is at perihelion (the point in its orbit closest to the sun) the planet will speed up by the discrepant 43 arc-seconds because the mass of the sun is warping space. A simple, two-dimensional experiment explains this concept. Imagine a flat sheet of rubber with a heavy ball bearing on it. The ball bearing represents the sun. Imagine a smaller ball bearing endlessly circling it—Mercury. At its closest approach to the larger ball bearing, the smaller ball will accelerate as it rolls down the incline created by the weight of the larger ball. Before rolling back out of the depression, then, the smaller ball has picked up speed, which advances the precession of its orbit around the larger ball. To Einstein's lasting credit, he used the orbit of Mercury as an acid-test for his theory. His calculations fit the observed 43 arc-second discrepancy perfectly, and the search for planet Vulcan was over.

## ADOPTING THE HARD ROCK ORPHAN

Although strange that Mercury, a planet so close to Earth, has been the object of only one mission, that oversight is about to be rectified. At the time of this writing, NASA's *Messenger* mission is en route to Mercury. *Messenger* (for MErcury Surface Space ENvironment, GEochemistry, and Ranging) launched from Cape Canaveral on August 3, 2004, and is already beyond the orbit of Venus. Like *Mariner 10*, the flight plan is complex, with two gravity assists using Earth and Venus. The payoff for this circuitous flight path will be no less than four encounters with Mercury, starting on January 14, 2008, and ending with *Messenger*'s arrival *in orbit* around Mercury on March 18, 2011. After that point, Mercury will join the other terrestrial planets in having a permanent, human-built spacecraft orbiting it.

Many of the questions *Messenger* will investigate are follow-ups to *Mariner 10*'s discoveries. Its investigations will be even more sophisticated, having been refined by three decades of comparative planetology—a discipline that did not exist when *Mariner 10* left Earth. When *Messenger* arrives in 2011, one of the first studies will examine Mercury's extreme density. *Messenger* will measure the abundance of elements in Mercury's surface rocks using its onboard X-ray and gamma-ray spectrometers. An elemental abundance similar to those of the other terrestrial planets would imply that the drag of the thin cloud of nebular gas at the dawn of the solar system favored accretion of dense particles into Mercury's core as the planets formed. Alternatively, the colossal heat from the sun might have vaporized the surface of Mercury, leaving it a charred cinder. In this case, the surface will be depleted of volatile elements like sodium and potassium. A third idea is that the merciless hail of meteorites during the Heavy Bombardment Period soon after the solar system's formation stripped away Mercury's newly formed crust and the upper layers of the mantle. If so, Mercury's surface will be composed of rocks depleted in elements like silicon, aluminum, and oxygen.

From the geologist's perspective, one of the most exciting aspects of the *Messenger* mission is that it will start illuminating Mercury's past. Earth's geological history has been studied for two centuries, Mars's geology has been under scrutiny for nearly two decades, and now even Venus's history is beginning to be interpreted. Given the paucity of information on Mercury's past, it is the geological orphan of the terrestrial planets. *Mariner 10* only imaged about 45 percent of Mercury's surface; thus, despite tantalizing glimpses of apparently unprecedented features, we have a very incomplete vision of the planet. With the imaging technologies of the twenty-first century, *Messenger* will help us visualize Mercury's past.

One of the crucial questions about Mercury's geological history relates to the plains that lap the Hermian mountains. As on the moon, such plains are likely remnants of ancient volcanism, and because they overrun the edges of certain craters, they must be younger than the craters. Since most of the craters were formed during the Heavy Bombardment Period of solar system history, the question becomes when, exactly, did volcanism on Mercury start, and when and why did it finish?

*Messenger*'s laser altimeter will tell us whether Mercury's core is liquid or solid. It will do so by measuring the planet's libration—its slow wobble around its rotational axis. If the core is still molten, calculations indicate that Mercury's libration will be twice as great as would be the case if the core is solid. By combining this data with information about the planet's gravity field, scientists will be able to determine the core's size.

*Messenger* will also investigate Mercury's enigmatic magnetic field. Since *Mariner 10* did not have time to fully characterize it, we still do not have a clear idea how strong Mercury's magnetic field is or if it has dipole-like characteristics. *Messenger*'s magnetometer will carefully study Mercury's field to establish its strength and variability.

Other instruments will measure the composition of Mercury's tenuous atmosphere and, most intriguingly, will investigate the mysterious material lurking in the eternal shadows at Mercury's poles. Is it water? Radar images taken from Earth indicate that, bizarrely, areas of Mercury's poles may be covered in patches of water ice where a combination of Mercury's strange rotational regime and deep shadows thrown by high polar cliffs prevent sunlight from hitting the surface. With the planet's extreme proximity to the sun, water on Mercury would be one of the biggest paradoxes in an already complicated solar system.

NASA is not alone in sending a new spacecraft to Mercury. In 2013, a joint mission from the European Space Agency and Japan named *BepiColombo* will blast off with a rendezvous with the planet scheduled for 2019. Bepi Colombo was one of the main architects of the original *Mariner 10* mission who calculated the orbit necessary for *Mariner 10* to make its three epic visits to Mercury. In a very real sense, this late Italian scientist is the godfather of Hermian science. *BepiColombo* will ask many of the same questions as *Messenger* but will have the advantage of more up-to-date instrumentation and a more favorable orbit. *BepiColombo* will also revisit Mercury's pivotal role in establishing Einstein's celebrated theory of general relativity by measuring the planet's perihelion acceleration, using more sensitive instruments. The mission will additionally study why spectroscopic observations do not reveal the presence of iron on the surface when for years it has been hypothesized that this element is a major constituent of the planet.

Mercury is now appreciated as a crucial "end member" in the spectrum of terrestrial planets. It is a jumble of extremes; it is the smallest, the densest, has one of the oldest surfaces, and exhibits the largest daily variations in surface temperature. Geologically, it is quite different from the three other terrestrial planets, yet it is paradoxically the least explored. Return visits to this strange world are long overdue, and it is a testament to this new golden age of interplanetary space exploration that Mercury is now so firmly on our "to-do" list.

# 3

# The Greenhouse in the Sky: Venus

Venus, the second planet from the sun, is named for the Roman goddess of beauty. To the Greeks, she was known as Aphrodite, and to the Babylonians, as Ishtar. All three are goddesses of love and splendor, an important clue to a singular property of Venus: besides the sun and moon, she is the brightest object in the sky, and as anyone knows who has seen her hanging like a limpid jewel in the twilight sky of dawn and dusk, she is indisputably dazzling. Yet Venus's allure is misleading. The swirling, lambent clouds that hide her surface and make her the brightest and most resplendent object in the sky also hide a dreadful secret: Venus is a world out of control.

The historical notion of Venus's habitability came from the observation that the planet's surface is hidden by a perpetual veil of cloud and from the beguiling notion that Venus is a tropical paradise because of its proximity to the sun. A Russian first discovered the planet's atmosphere, which may explain the Soviet obsession with

Venus during the latter half of the twentieth century. In 1761, at the University Observatory in St. Petersburg, Mikhail Vasilievich Lomonosov, who has been referred to as the father of Russian science, studied Venus during a rare transit of the planet across the solar disk. As Lomonosov attempted to determine Venus's diameter, he was stymied by an odd fuzziness at the edges. After pondering this mystery, Lomonosov came to an amazing conclusion: Venus had an atmosphere! He had no idea what fertile territory this innocent observation would open up, for with his discovery of Venusian clouds, the myth that Venus might be a habitable paradise was born.

During the late nineteenth century, the possibility of life on other planets was not only considered possible, it was deemed overwhelmingly likely. This idea arose in the aftermath of the Copernican revolution in the sixteenth century and became known as pluralism, or the belief in the existence of life on an infinite number of habitable planets throughout the universe. As with any new scientific theory, opinions ranged from complete disagreement—life cannot exist elsewhere in the universe—to the extreme position that the default condition of all planets is to harbor life. The eighteenth and nineteenth centuries were the heydays of pluralism. Early in the nineteenth century, pluralism was popularized in part by the pugilistic intellectual dueling of David Brewster and William Whewell. Brewster, a Scottish minister and scientist, was a staunch supporter, whereas Whewell, master of Trinity College, Cambridge, came to vehemently reject the idea.

By any standards, though, the mid- to late nineteenth century witnessed deep and controversial thinking about the possibility of extraterrestrial life. The reason for this extreme extraterrestrial fervor can be summarized in a single word: Darwin. It is impossible to underestimate the impact *On the Origin of Species by Means of Natural Selection* had on the Victorian public. It was a talking point the length and breadth of the empire, in salons, taverns, and gentlemen's clubs as well as in the learned societies that attracted not only scientists but the interested lay public. Even eminent politicians such as Gladstone and Disraeli were frequent visitors to groups such as the Geological Society of London. By defining the mechanism for the formation of new species, Darwin legitimized biology and, for pluralists, sowed the obvious seed that, if organisms can develop and evolve on Earth, then why not on other planets? Darwin's theories laid the groundwork for what we regard today as the new science of astrobiology.

The British astronomer and writer Richard Proctor was an ardent supporter of pluralism. His extremism did not lead him to believe that all planets are inhabited, however—far from it. He regarded Mercury as incapable of supporting life because its proximity to the sun would expose any life forms to lethal levels of solar radiation. He theorized that the gas giants Jupiter and Saturn would be too hot to support life despite the theories of Brewster and Whewell that suggested the opposite. Venus, however, was the planet most likely to support life. Proctor wrote:

> It is clear that, merely in the greater proximity of Venus to the sun, there is little to render at least the large portion of her surface uninhabitable by such beings as exist upon our earth. This undoubtedly would render [the sun's] heat almost unbearable in the equatorial regions of Venus, but in her temperate and subarctic regions a climate which we should find well suited to our requirements might very well exist. . . . I can find no reason . . . for denying that she may be considered the abode of creatures as far advanced in the scale of creation as any which exist upon the earth.

Everyone tacitly assumed that Venus's clouds, like Earth's, were composed of water vapor; this assumption, along with Venus's proximity to the sun, made the idea of the planet as a tropical swamp world hard to avoid. Given this reasoning, notions about the habitability of Venus might seem logical even now.

By the end of the nineteenth and into the twentieth centuries, scientists were casting about for a specific vision of what Venus could be like. In the absence of other alternatives, they turned to Earth's own long and diverse history. Before long, they turned up trump cards. The geological period known as the Carboniferous occurred toward the end of the Paleozoic era. During this period, 345 to 290 million years ago, the continents of the world were gathered along the equator into a single landmass known as Pangaea. This time was also one of much evolutionary activity as the great engine known as natural selection favored flora and fauna of immense size. The period produced ferns and primitive gymnosperms that towered hundreds of feet into the air. Animals of the Carboniferous period boasted the largest insects ever known, including huge dragonflies. By the early twentieth century, the Carboniferous era was widely perceived as the ultimate swamp world and, by a curious leap of the imagination, became the predominant analogy for understanding Venus. This concept was curious because a moment's thought shows that the notion does not bear close scrutiny. There is no evidence that the Earth of the Carboniferous period was wreathed in mist and clouds any more often than today's Florida Everglades. Why should Venus's clouds hide a Carboniferous swamp world? It doesn't make sense.

The idea of Venus as a swamp world was seriously advanced in 1918 by Svante Arrhenius, a scientist who had won the 1903 Nobel Prize for his work on electrolysis. Arrhenius's equation, which quantifies the temperature dependence of many chemical reactions, underpins all of modern chemistry, so the man had excellent scientific credentials. In his book, *The Destinies of the Stars*, Arrhenius wrote:

> Everything on Venus is dripping wet—a very great part of the surface is no doubt covered by swamps corresponding to those on earth in which the coal deposits were formed. . . . The constantly uniform climatic conditions which exist everywhere result in an entire absence of adaptation to changing exterior conditions. Only low forms of life are therefore represented, mostly no doubt belonging to the vegetable kingdom, and the organisms are nearly of the same kind all over the planet.

Arrhenius's writing is compelling and has, at least superficially, a certain kind of logic. Implicit in his statement are two vital concepts: the nebular hypothesis of Laplace and the Darwinian idea that environmental change drives evolution.

Pierre-Simon de Laplace was born in 1749. His most notable contribution to astronomy was to place the ideas of Isaac Newton on a firm mathematical footing. It might seem surprising that the most famous mathematician of all time needed somebody else to help him get his sums right! But although Newton used geometry to formulate his rules of celestial mechanics, Laplace was the person who converted them into calculus, the branch of mathematics that deals with the *rate of change* of quantities. A broad thinker, Laplace also proved that the solar system was held in a steady state by the dynamic forces continuously operating upon it and explained why the orbital speeds of Jupiter, Saturn, and the moon vary over long timescales.

Laplace's most famous and arguably most valuable contribution is what Arrhenius obliquely refers to above: the nebular hypothesis. Laplace suggested that the planets of the solar system condensed out of gradually cooling concentric rings of gas left over from the solar system's formation. The rings furthest from the sun cooled and coalesced first, with the inner planets being formed later. The hypothesis thus implies a gradient of increasing planetary age as one moves outward from the sun: Mercury is younger than Venus, which is younger than Earth, and so on, with the oldest planets at the farthest reaches of the solar system. This idea led to the concept of planetary senescence, which became very influential to those thinking about planetary life in the nineteenth and twentieth centuries. Planetary senescence viewed Mars as a planet past its prime, whereas Venus was a world still struggling through the primordial soup phase that Earth passed through during its Carboniferous period.

Darwin was the other influence informing Arrhenius's ideas. Darwin, as I discuss in my book *The Silent Landscape*, explicitly linked evolution—powered by natural selection—to environmental change. Changes in the environment force organisms to adapt, resulting in more complex life forms. Scientists of the late nineteenth century had a simplistic understanding of evolution. For example, Wyville Thomson and his "Scientifics" fully expected to find "living fossils" in the depths of the ocean during their 1872–1876 expedition to dredge the depths of the sea. Their logic, based on the misconception that evolution proceeds at a significant rate only in environments that vary substantially and rapidly, reasoned that the unchanging landscape of the deep ocean would harbor atavistic life forms left behind long ago by the rapidly evolving sunlit world on land. We now know this theory is a vast oversimplification and that the process of evolution on our planet, let alone elsewhere, is controlled by the interaction of complex forces we have barely begun to comprehend. However, Arrhenius was himself a product of his environment, and, not for the first time, we see how science reflects the prevailing intellectual culture of its time.

Once we understand the background to Arrhenius's ideas, it is easy to see where the early twentieth-century vision of planetary progression came from. To late nineteenth-century and early twentieth-century biologists, evolution proceeded along more-or-less straight lines from simplicity to complexity as a function of adaptation to changing environmental conditions. All evolutionary theorists of the time, dating back to Darwin himself, believed in this progression. If evolution proceeded in straight lines on the Earth, why not extend this idea to our planetary neighbors?

The idea of evolutionary progression is simple and seductive, and despite the rages of intellectual giants like Stephen Jay Gould and Richard Dawkins, it has been very hard to dislodge this notion from the popular consciousness. Even today, many members of the public believe that evolution means the development of more complexity from simplicity. It seems self-evident; the oldest and most primitive organisms on Earth are bacteria and, before them, naked strands of replicating nucleic acids (the so-called RNA world). Today, the most complicated organisms on Earth are bipedal ape descendents with handheld computers and brains capable of composing piano concertos. Since progression on Earth is so believable, it is hard not to export the idea off-world.

But evolution has not proceeded in straight lines—many of the dinosaurs were certainly highly sentient, and certain groups such as the raptors had social organization. Vertebrates originally emerged from the sea about 400 million years ago, and the development of complex life since then has been largely concentrated in the tetrapods—the land-dwelling vertebrates. It is now known that whales evolved from land-dwelling animals related to turtles and *went back* to the sea. In other words, the sea was *recolonized* by tetrapod descendents—the whales and dolphins. Progress is not unidirectional on Earth, which renders the concept's application beyond our planet difficult at best. In addition, because Laplace's notion of solar system formation has been replaced, as we will see, with more modern ideas, there is even less reason the planets should conform to this notion of progress. Yet the notion of evolutionary progress held sway for much of the twentieth century, as did the idea that Venus is an up-and-coming world, steadily making its way through some kind of extraplanetary Carboniferous period.

By the second decade of the twentieth century, Arrhenius's ideas about Venus's "Carboniferous Time-Warp" nature had settled into the popular consciousness. But it was not long before doubts arose. Only a few years after Arrhenius's 1918 paper, Venus's clouds became the subject of a new science, astronomical spectroscopy, which analyzes the chemical composition of remote planetary atmospheres by measuring the amount of absorbed light. By the 1920s, it was becoming clear that spectroscopic observations were not showing much water vapor in the Venusian atmosphere. They were, however, revealing a great deal of carbon dioxide. It was the first concrete evidence that the swamp-world concept of Venus was wrong.

Nonetheless, scientists clung to the Carboniferous version of Venus in the face of mounting evidence to the contrary. Even maverick British astronomer Fred Hoyle of Stonehenge fame suggested an alternative to reconcile the ideas of swamp-world Venus with the atmosphere's newly discovered high carbon dioxide levels. Hoyle suggested that the surface of Venus might be covered in an ocean of oil and that the atmosphere might be composed entirely of oil droplets, thus accounting for the spectroscopic measurements. But such a world would not be a Carboniferous swamp world, for oil and coal are not formed in the same manner, and most of our world's coal formed during the Carboniferous period. Oil formed on our planet during a series of geological periods ranging from the Precambrian to the Tertiary—a much greater span of geological time than the Carboniferous—through the decomposition of (usually marine) algae into kerogen and then to oil. Coal, by contrast, formed by the decomposition of land plants on the margins of swamps and seas. Thus the presence of oil does not imply the presence of coal. Despite Hoyle's efforts, and those of other scientists, the growing evidence dispelled the concept of a water vapor–bearing atmosphere on Venus.

The concept of a swamp-world Venus enjoyed a brief renaissance in 1955 when the astronomers Frank Whipple and Donald Menzel argued that the planet's atmosphere might be rich in frozen water—

specifically ice crystals—that would not show up spectroscopically. They suggested that Venus was completely covered in a world-girdling super ocean that might be partially frozen. Any land mass above the sea surface would remove the carbon dioxide from the atmosphere and fix it as solid carbonate (equivalent to the limestones and chalks of our own world). This theory paved the way for a refinement of the idea of Venus as a throwback to our own geological past. A watery Venus (as opposed to a swampy Venus) would be more like the Earth of the Cambrian period (about 500 million years ago) than the Earth of the Carboniferous period. The Cambrian was a time of exclusively marine organisms, and Venus's association with it gave birth to the myth of the oceans of Venus.

The argument over whether Venus was some Carboniferous or Cambrian paradise inched toward resolution soon after the work of Whipple and Menzel. In 1956, radio observations of Venus showed that the planet was mysteriously emitting massive amounts of microwave radiation. What could account for this emission? Some researchers suggested that the planet was acting like a mobile phone on steroids because the surface of the planet was very, very hot. Clearly, a hot surface was incompatible with a frozen sea-world Venus. Most unusually for questions in comparative planetology, the answer was about to be found in a simple way.

## HOT VENUS

The dawn of the space race sported two main players: the Soviets and the Americans. These two competitors were fighting for three important targets: the moon, Mars, and Venus. As we saw in Chapter 2, the first probe sent to another planet, *Mariner 2*, was an American attempt to investigate Venus. *Mariner 2* was launched on August 27, 1962, and by any standards, the mission was an unparalleled success. The tiny craft carried a microwave radiometer specifically to answer the loaded question of whether Venus was a subtropical paradise clawing its way up the evolutionary scale or something you could cook the Sunday roast in.

Technically, what the *Mariner 2* scientists were looking for was a phenomenon known as "limb darkening." If the microwave emissions came from the edges of the planet (where the atmosphere as seen from an approaching spacecraft is the thickest), then the atmosphere would be a "bright" microwave emitter. This result would imply that the surface might be at least relatively cool, and Venus could retain its status as a mysterious, cloud-shrouded water or swamp world where all possibilities, including the existence of life, could remain unchallenged. The vision of Venus promoted by science fiction writers like Robert Heinlein, John Wyndham, and Isaac Asimov would remain plausible. Conversely, if the microwave radiometer detected a peak in the amount of radiation from the *center* of the planet, where the atmosphere would be thinnest, then the surface must be hot, and life, unlikely.

On December 14, 1962, *Mariner 2* grazed Venus, skimming past the planet at a distance of only 21,000 miles, with the microwave radiometer working perfectly. The results were unequivocal: the limbs of the planet were dark. The surface of Venus is thus hot. In fact, the results showed that the surface of Venus is not just hot, it is as hot as the interior of a self-cleaning oven. *Mariner 2*'s instruments showed that the surface of Venus is about 500 degrees C—no global oceans, no swamps, no giant tree ferns, no enormous insects, and no amphibian-like creatures crawling their way toward sentience.

The (literally) burning question became, how could Venus be so different from its nearest neighbor, Earth? By finally managing to penetrate the veil of Venus's clouds, we had discovered that the surface, far from being able to support life, is sufficiently hostile to make all known forms of life extinct. It was an important step in our understanding of Venus, for we had finally managed to gather some data about the planet's surface. Yet the mystery, far from being solved, had deepened dramatically.

One immediate effect of the news from *Mariner 2* was that America lost all interest in Venus and shifted its focus to Mars, the moon, and deep space. In startling contrast, the Soviet focus on Venus intensified. The Soviet interest may have stemmed from the days of Mikhail Vasilievich Lomonosov, or perhaps it was simply that a planet with a surface temperature of 500 degrees C seemed attractive to scientists and engineers accustomed to working in the long, dark communist winter of Stalin and Khrushchev (not to mention the more literal winters of the Eurasian steppes). In any case, the message from the Central Committee was clear: The Soviet space establishment was to focus the attention of its nascent unmanned space program on this nearby, bright planet that glimmers so temptingly in the evening sky, with the goal of landing a probe on it. Such an order was nothing if not audacious for, at this time in the early 1960s, no one had even landed a probe on the moon. Nonetheless, thus was born the Venera program.

## IRON MEN OF THE EAST

By the beginning of the 1970s, the Soviet Union was losing its lead in space to the United States. The Americans had successfully landed a man on the moon, and its earliest Mars probes were successfully sending back images of the red planet. By contrast, a concerted Soviet campaign to land probes on Mars had spectacularly failed, and the pressure was on to produce results. Also, it was known that the Americans were planning to spend a billion dollars on two probes—the *Vikings*—that were to land on the red planet to celebrate the bicentennial of the United States. These probes would carry apparatus enabling them to search for life.

At the end of 1973, Roald Sagdeev, newly appointed director of the IKI, the Soviet Space Research Institute, decided that the time had come to cut Soviet losses. Although two Soviet Mars probes were expected to reach the planet in February and March of 1974, the Soviet record had been so dismal that Sagdeev lacked confidence that these missions would succeed. With launch windows only every two years—and the next one to be used by the Americans to dispatch their *Vikings*—there was little hope the Soviets could overtake the Americans in their exploration of Mars, so Sagdeev decided to target Venus instead.

The Soviets had had some success with previous Venus probes, but Sagdeev wanted to up the ante. With the newly developed Proton booster, spacecraft weighing more than five tons could be placed in orbit; hence suitably armored craft could be sent to Venus and hopefully survive the crushing pressure and extreme temperatures there. The launch windows were more frequent, and the transit time to Venus was faster than to Mars. Transit time was particularly important because the quality of Soviet onboard electronics was poor; the quicker the travel period, the less chance of something going wrong. If they could just return pictures, the first photos ever from the surface of another world, they would beat the Americans—regardless of their over-hyped technology, massive budgets, and experiment-festooned Mars landers.

The new Venera spacecraft were completely redesigned from the previous generation. They were sternly beautiful in a supremely Soviet way, radiating brute force and a grim determination to get the job done in equal measure. Each of the newly redesigned Venera landers stood over two meters tall, with a span of almost seven meters wide when the solar panels were unfurled.

*Venera 9* was launched atop a massive Proton rocket from Site 81 at the Baikonur Cosmodrome on the Kazakhstan steppe on June 8, 1975. *Venera 10* followed six days later. *Venera 9* entered the Venusian atmosphere on October 22, 1975. It had been a nail-biting four months for the IKI scientists and engineers from NPO Lavochkin (the Soviet aircraft engineering firm), but on October 22 *Venera 9* had become the first spacecraft to place a satellite in permanent orbit around another planet.

*Venera 9*'s lander approached the cloud-wreathed surface of Venus at a staggering 10.7 kilometers per second. At 125 kilometers above the surface, it encountered the outer layers of atmosphere. The lander had a cylindrical upper portion that looked like a top hat, and a ring around the lower portion that resembled a miniskirt. The top hat served several purposes: inside was a spiral cylinder that was both the parachute housing and the high-gain antenna; the miniskirt was actually an aerobrake to slow the craft down after parachute deployment. Below the miniskirt, the lander splayed out into a trampoline-like ring—the landing amortization ring—which would cushion the impact of the craft's seven-meter-per-second descent. The first generation of Venera probes had discovered that the atmosphere of Venus was sufficiently dense that the final part of the descent could be made without the aid of parachutes. This was an advantage because parachutes would have kept the landers in Venus's hostile atmosphere too long, and with an expected lifetime for the lander of only thirty minutes, such a delay risked failure of the landing phase. And failure was not an option—the Soviet goal was to be the first to take pictures on the surface of another planet.

After the spacecraft completed its initial aerodynamic breaking, the parachute compartments jettisoned at sixty-five kilometers, and the small pullout parachute began streaming up into the Venusian atmosphere as the top half of the spherical re-entry shell blew away, slowing the craft to around 150 meters a second. At sixty-two kilometers, the three main parachutes opened, while four seconds later, the lower hemisphere of the landing shell separated, falling away into the murk under its own weight. *Venera 9* was now revealed in her full glory. The craft spent the next twenty minutes descending through the thick coils of Venusian atmosphere, transmitting back to Earth an untold wealth of new data from the lander's comprehensive instrument packages. Apart from a thermometer and barometer to measure the extreme temperature and crushing pressure of that cloud-shrouded world, each of the new *Venera* landers carried accelerometers to measure the

G forces as the probes plunged into the Venusian atmosphere, mass spectrometers to determine the molecular makeup of the atmosphere at altitudes from sixty-three to thirty-four kilometers, and nephelometers to measure the atmospheric density. And when the landers touched down, on each there was a multichannel gamma spectrometer designed to test for the presence of naturally occurring radioactive elements in the soil, a densitometer to measure the soil density, and perhaps most extraordinary of all—the instrument that was the heart of this mission—the telephotometer or panoramic imaging system. Each lander had two photographic scanning devices whose sensors—secure behind quartz windows and using a complicated system of hinged mirrors—would take the first pictures of the surface of another world. Each imaging system was located in a gondola about ninety centimeters above the base of the lander, and each camera could provide images with a resolution of about 70,000 pixels. Each pixel would then be encoded with a seven-digit number that would be transmitted to the orbiter and, from there, to Earth. To enable the cameras to see, each lander was equipped with two powerful floodlamps, for previous Venera missions had shown that the best light levels on Venus were like a foggy dusk in Moscow.

The first data from the descending spacecraft began to stream into the room on schedule. Engineers and technicians were leaping and yelling, pounding each other on the back in celebration. Then, again right on schedule, at fifty kilometers the main parachutes jettisoned, and the lander accelerated as it went into freefall. The miniskirt worked perfectly, the ring around the now-naked lander preventing it from accelerating too fast. Gradually the lander slowed again as the density of Venus's atmosphere increased with decreasing altitude. For the scientists and engineers, it was an extraordinary feeling to know that a seven-ton, Soviet-built Iron Man was being slowed only by the turgid atmosphere of a hostile world. Had Soviet engineering finally met its match in the atmosphere of Venus?

As it approached the surface, *Venera 9*'s instruments showed that the shrieking winds that had buffeted it between twenty and forty kilometers were decreasing. And then, at ten kilometers, something very strange happened—wind speeds decreased as though a switch had been thrown. As *Venera 9* drifted lower, it was as though it were approaching the bottom of some deep ocean canyon.

At 8:16 AM Moscow time, the altimeter display stopped counting down. Allowing for the three-minute time delay between the orbiting bus and Earth, *Venera 9* had touched down at 8:13 AM. Its speed at touchdown had been only seven meters per second. The gas-filled landing amortization ring cushioned the impact so well that the spacecraft remained upright. The Soviets had touched the surface of Venus five years earlier with *Venera 7*, but on that occasion the event was marked only by the stilling of the altimeter. But this time was different. This time, pictures were to be taken.

The covers protecting the quartz windows over the camera compartments were to jettison immediately upon landing. The two cameras were configured together to provide a 360-degree view of the landscape around the spacecraft. The scientists waited an agonizing three minutes, and then, on the screen in front of them, pixel by pixel, a grainy image began to emerge.

As the lines of the image coalesced top-to-bottom on the screen, the first thing they noticed was the extreme curvature of the image. That curvature, however, was a function of the nodding mirror arrangement that had been placed in front of the photomultiplier tube. To get the entire 360 degrees with only two cameras, each camera had to cover 180 degrees, and as a result, the image was curved. But only one-half of the screen was filling with data. The other half—the data from the other camera—remained resolutely blank. One camera, for whatever reason, was not working.

But by now the image from the functioning camera had built up sufficiently to give them their first view of the surface of another world: a monochrome image that showed a grey plane littered with variously sized eroded boulders and flattened pebbles, lying in a fine-grained matrix of dust-sized particles. And there, at the bottom of the screen, an image began to emerge that was not of Venus—a regular shape in an alien world of irregularity, a curved face of metal with a familiar checkerboard hatching. The smooth, hard curve of the landing amortization ring—that is, *Venera 9* itself—resting solidly on the surface of Venus.

By any standards, the Soviet Venera program was a staggering success and rebuts any suggestion that the Soviet space effort ran out of steam as soon as the Americans got their act together. After a few early failures (that the Soviets declined to publicly acknowledge), *Venera* 4 had arrived at the shrouded planet in 1967, eight years before *Venera* 9's first successful pictures, and returned the first data about its atmosphere. This data helped put to rest, once and for all, any notion of Venus as a sea or swamp world. The Soviet Union used just about every feasible conjunction—favorable launch windows—to send spacecraft to Venus from the 1960s until the middle of the 1980s.

In 1969, *Venera 5* and *6* were also successful atmospheric probes, and in 1970, *Venera 7* became the first Earth-built spacecraft to land on the surface of another planet. *Venera 8*, in 1970, measured wind speeds through Venus's thick atmosphere. Then in 1975, having missed the 1973 launch windows to concentrate on the construction of their new generation of probes, the Soviet Union landed *Venera 9* and *10* and returned the first pictures from another planet's surface.

Three days after *Venera 9*'s successful landing, on October 25, 1975, *Venera 10* entered the Venusian atmosphere. It touched down a little over two thousand kilometers due south of its sister craft. Like *Venera 9, Venera 10* performed flawlessly during the descent and landing phases of the mission, and as with *Venera 9*, one of the two lens caps failed to come off. The images *Venera 10* sent back of its landing site showed an almost perfectly flat, scoured-looking landscape with virtually no topography. The spacecraft had apparently come to rest on the flattened surface of an enormous boulder in the middle of a Venusian desert.

Perhaps the most impressive achievements of *Venera 9* and *10* were their longevity. Each craft had been designed to withstand conditions on Venus's surface, where temperatures can reach 500 degrees C and the pressure is equivalent to being three thousand feet underwater on Earth, for thirty minutes. *Venera 9*, however, functioned for fifty-three minutes, and *Venera 10*, for a little over an hour.

Buoyed by its success, the Soviet Union continued to send probes to Venus. *Venera 11* and *12* suffered from even more acute imaging problems than their predecessors when both lens caps failed to detach and no pictures were retrieved. *Venera 13* and *14* in 1982, however, were successes, returning the first color pictures of the surface. *Venera 14* suffered ignominy when one of its camera lens caps—while successfully detaching—landed on the ground exactly where the soil densitometer would deploy. Presumably they are still there: a complex instrument for measuring the density of alien soil lying atop an artifact from a nation that no longer exists, the last legacy of a multimillion-ruble mission.

Any problem with seeing Venus was triumphantly made irrelevant when *Venera 15* and *16*—the last in the series—initiated what was to become one of the most important imaging initiatives directed at the shrouded planet: radar mapping of the surface of Venus. Once again, the Soviets led the way, and the Americans followed in the 1990s with comprehensive radar mapping of the surface of Venus using their Magellan probe. In addition, in 1985 scientists from NPO Lavochkin were the first to deploy balloons in the Venusian atmosphere from the *Vega 1* and *2* probes, which had been dispatched to rendezvous with Halley's Comet. The balloons successfully measured wind direction in Venus's upper atmosphere.

Extraordinarily, all of this activity was orchestrated by the Soviet Union in the final years before it collapsed under the weight of its own bureaucracy and the colossal expense of keeping pace economically with Western capitalism.

## GETTING BACK IN THE VENUS GAME

It is not clear why the Americans abandoned Venus and left it to the Soviet Union to explore. Perhaps the answer is because the early years of the space race were fuelled not as much by the needs of science as by Cold War paranoia. John F. Kennedy's mandate during his 1962 landmark speech clearly made placing a man on the moon the immediate objective of the American space program. The exorbitant expense of landing Americans on the moon—$24 billion at 1960s prices—not only destroyed the possibility of going to Mars but eventually curtailed the moon landing program itself.

During the early 1970s, the American emphasis shifted from landing humans to landing machines on extraterrestrial surfaces. By the late 1970s, America had returned its attention to Venus. With the favorable conjunction between Venus and Earth in December 1978, the skies around the shrouded planet became positively crowded as the Americans sent the Pioneer Venus mission, consisting of a bus carrying no less than four probes designed to explore the Venusian atmosphere.

America's renewed interest in Venus was arguably a result of the spectacular successes of the Soviet Venera program. After all, the Soviets had managed to land the first manmade object on the surface of another planet and had returned the first pictures of that surface. The Venera program had shown just how alien other worlds could be. Clearly the structure of Venus's atmosphere was highly complex, with cloud layers of differing densities and chemical compositions at different altitudes. But there were many questions left unanswered. For example, did the cloud layers discovered by the *Venera*s extend uniformly around the entire planet, or did they vary as earthly clouds do? And, crucially, there was an emerging awareness that the Venus we were coming to know in the 1970s might not be representative of Venus as she had once been. As long ago as 1969, Carl Sagan suggested that the cause of Venus's extreme heat could be the result of a runaway greenhouse effect. This theory led to more questions: If Venus was an out-of-control world, what had triggered the loss of control in the first place, and when *exactly* had that happened? Had there ever been water on Venus? If there were no seas or swamps *now,* had there once been a time when they had existed? Had Venus always broiled, or had that searing planet once been cooler, perhaps as cool as Earth? And underlying these questions was that nagging anxiety: If it had happened once, could it happen again, perhaps someplace else? In short, could it happen to us?

America's *Pioneer Venus* probe was the first attempt to answer questions like these. Following the Soviet initiative, *Pioneer Venus* was composed of two main spacecraft—an orbiter carrying a radar mapping altimeter (to add to the topographic data already being collected by the Soviets) and a bus carrying four probes that would descend to the surface. The large entry probe would take gas samples to measure the composition of the atmosphere. The other three smaller probes were *sans* parachutes and carried instrumentation designed to measure light levels during their descent. The instrument sensors were protected by windows of sapphire and diamond.

The small probes were named *Day*, *Night*, and *North* for self-explanatory reasons: one for the day side of the planet, one for the night side, and one for the largely unexplored northern region of the planet. They entered the atmosphere about ten minutes after the large probe and immediately started sampling the clouds. Meanwhile, back on Earth, mission controllers realized that the large probe had a problem. A tiny droplet of the Venusian atmosphere had entered the mass spectrometer sampling tube and lodged in front of the sensor. Instead of sampling the continuously changing Venusian atmosphere as the probe descended, the sensor continuously sampled only the single drop that it had collected high in the atmosphere. It was America's first taste of the kinds of problems that had plagued the Soviets.

Eighteen minutes after entering the Venusian atmosphere and twenty-eight miles above the surface, the large probe abandoned its parachute and, like its Soviet forebears, continued in free fall through the increasingly dense atmosphere. Thirty-six minutes later, having sampled continuously all the way, it crashed into the baking plains of Venus and fell silent. The other three simpler probes all performed successfully. Each impacted the surface, and one—*Day*—continued functioning and sending back information for another sixty-eight minutes. It failed when its internal temperature reached 260 degrees F.

One of the intriguing mysteries surrounding the Pioneer Venus expedition, one that remains unsolved to this day, is why all four probes went haywire at an altitude of about 12.5 kilometers. At that altitude, a simultaneous power spike jolted the onboard instrumentation; the temperature and pressure sensors transmitted weird numbers, and some instruments ceased functioning completely. In 1993, over a decade after the Pioneer Venus expedition, NASA convened a conference to investigate the reasons for this "12.5-km anomaly," but no clear answer emerged.

The bus that had carried the four probes then plunged into the Venusian atmosphere itself. Without a heat shield for protection, it was never intended to last long. Even so, in the two minutes before its immolation high above Venus's southern hemisphere, it managed to radio home valuable measurements about Venus's upper atmosphere and ionosphere, regions never before probed.

Of the six spacecraft America dispatched on this single expedition, only the orbiter was left, returning data to Earth at a more leisurely pace since it did not have to anticipate its own destruction. The orbiter carried instruments to measure cloud-top wind speeds to complement the measurements made by the entry probes at more varied altitudes. More importantly, the orbiter was equipped with a synthetic aperture radar (SAR). This aspect of the Pioneer Venus mission was a departure from all previous missions to the shrouded planet, because the SAR's function was to map the surface of the planet. Previously, the focus had been on understanding Venus's atmosphere, and the surface of the planet had been largely ignored. With each orbit, the spacecraft's radar returned a narrow image of the surface, and back on Earth, the strips were painstakingly reassembled into a global topographic map of our nearest planetary neighbor. The mapping of Venus had begun.

## CREATED IN MY MAGIC EYE

The results of the mapping efforts were spectacular. For the first time, Venus was revealed as a recognizable place. Whereas Venus was previously known only from snapshots sent back by dying Venera probes or as a distant, speeding object fleetingly glimpsed from deep-space probes passing by, she was now being minutely examined by an observer who was there to stay for awhile. (Incredibly, the *Pioneer Venus* orbiter was still functioning fourteen years later in 1992.) The spacecraft's SAR revealed a planet with topographic highs and lows—a planet, in short, much like the Earth. In fact, the conclusion was immediate and inescapable: Venus looked much more like Earth than anyone had expected.

The *Pioneer Venus* imaging of Venus's surface put the science of comparative planetology on a new footing because it allowed a new series of questions to be asked. The most immediate of these was, if the surface of Venus looks like that of Earth, does that mean it was formed by the same types of processes? In other words, does Venus have tectonics?

This kind of question could not even have been asked only twenty years earlier. Only in the 1960s did a majority of geologists accept that the continents of our world moved around, bumping, grinding, and generally crashing into each other as they ploughed their stately way through an ocean of magma. Only then was it generally accepted that the entire history of Earth—all 4.5 billion years of it—had been characterized by the continents moving together, forming large supercontinents, then breaking apart and forming smaller continents.

When the Pioneer Venus scientists examined the first map of the surface of Venus, they immediately applied the hot new theory of plate tectonics to a radically different world. Some pointed to a ring of topographic highs that girdled the equator of Venus. These included what became known as Aphrodite Terra, a huge area—about the size of Africa—that rises two miles above the surrounding plains. To the east of Aphrodite Terra is another highland region called Beta Regio. Even further east is another series of small highlands called Eistla Regio. Scientists suggested that this system of apparently interlinked highlands around the midriff of Venus is equivalent to the plate tectonic ridges of our own planet, where new crust is formed. The analogy with Earth was further extended by the discovery of an enormous and very mountainous highland region in the northern hemisphere—Ishtar Terra—that they suggested was equivalent to an Earth continent. Ishtar Terra, they argued, was an area of low-density rock pushed up by tectonic plate motions that had started at the equatorial spreading center—the Venusian equivalent of the Himalayas.

There is, however, one aspect in which Venus is unarguably similar to planet Earth; she has very few impact craters. To understand the significance of this fact, one has to understand the central role craters play for the average comparative planetologist. Craters are the hallmarks of eternity, produced by the collision of asteroids and meteorites with a planet's surface. These collisions fortunately are rare, but given enough time, their spoor, craters, will accumulate on the surface of any planet or moon that does not have a resurfacing mechanism like, for example, Earth's tectonics. The more heavily cratered a surface, the older it is; studying craters on the face of a planet is like watching the moving hands on the face of a clock.

Our own moon is covered with a dense layer of overlapping craters, many big enough to be easily seen with the naked eye from Earth. Mercury, as we have seen, is similarly pockmarked. Both of these bodies have ancient, scarred surfaces testifying that, on both, tectonics is dead or, more correctly, never started in the first place, and the craters provide insight into their ages.

Conversely, Earth has relatively few craters because the surface of our planet is continually resurfaced, albeit at a rather stately rate. The *Pioneer Venus* orbiter showed *virtually no craters at all.* Using the crater-aging technique (that is, that craters accumulate at a stochastically random rate), scientists calculated the surface of Venus at no older than 500 million years. In a human timescale, this may seem a long time, but when measured against the age of the solar system, a staggering 4.5 billion years old, 500 million seems rather insignificant. Against this timescale, Venus's surface was repaved only yesterday. And this relatively modern resurfacing appears to have occurred at the same time globally.

This truly astounding finding highlights a major difference between the tectonic regimes of Venus and Earth. The Earth is more-or-less continuously resurfaced—a process that happens all the time. The twitches and rumbles of this tectonic conveyor belt are represented on Earth by the occasional earthquake or volcano. Such events are usually catastrophic at the time—just ask any survivor of the 2004 Boxing Day tsunami. Arguably, however, these tremblers and eruptions are the release of steam in a continuous and controlled fashion. Not so on Venus; apparently the Venusian tectonic forces build to a boiling point and then blow at every seam, all at once.

This appears to have been what happened 500 million years ago during the last great Venusian resurfacing event. On Earth, another event—no less important in its own way to us—was happening at about the same time. This was the period geology knows as the Cambrian period, when the complex animals who are our direct ancestors appeared, by the standards of geological time, overnight. Because it happened so quickly (in less than 10 million years), this event is known as the "Cambrian explosion." At the same time, then, that life on Earth started down the road to spacecraft-building apes, Venus was undergoing a tectonic upheaval so dramatic it literally scoured the face of the planet clean.

## OTHER DAYS, OTHER EYES

The radar eyes of *Pioneer Venus*, augmented by data from the *Venera 15* and *16* probes, confirmed ground-based radio telescope data about the nature of the Venusian surface that had been gradually accumulating since the 1950s. The ground-based observations alone had been insufficient to understand the face of Venus; it was the Pioneer Venus mission that solidified suspicion into scientific theory. As always in science, more questions were raised by the success of Pioneer Venus than were answered, and it was not long before those studying Venus wanted to go back with better eyes to improve the resolution of the radar mapping.

Unfortunately, the age of gold-plated, money-is-no-object space exploration was coming to an end. The colossal cost of the Apollo moon landings had finally come home to roost with the American taxpayer. The Mars Viking missions and the Voyager missions to the outer planets had been dispatched, and although the Voyager missions would be spectacularly successful, the Viking missions spectacularly failed. With the end of the space-race era, the immense pressure to succeed in space that had dominated during the 1950s and 1960s was no longer evident. The 1980s were the era of Reagan, Gorbachev, and glasnost; the Cold War was thawing significantly. What was the point of fighting each other for control of distant frontiers when even local frontiers were losing their meaning? The second great era of solar system exploration was not necessarily coming to an end, but mission planners would now have to keep an eye on the budget as well as the course trajectory.

As early as 1981, a Venus orbiter carrying a high-definition radar camera was proposed. Named *VOIR* for *Venus Orbiting Imaging Radar*, not only would it carry high-definition radar to map the surface at extremely high resolution, it would also carry a comprehensive array of experiments to study the Venusian atmosphere in more detail. Conceived in the white-hot energy of the Viking and Voyager era, it was to be a well-equipped expedition with a heretofore unseen array of scientific instrumentation. Without the spur of the Evil Empire's space-based super-bogeymen, however, Congress canned the mission, and its $800 million price tag was axed from Ronald Reagan's 1982 budget.

By now, planetary scientists knew that the only way to study planets effectively was with spacecraft. They did not consider them toys or symbols of national prestige; rather, they were the necessary instruments of science. If $800 million was out of the question, could a cheaper mission prevail? After all, the radar mapping device—required to map the Venusian surface in detail—was the centerpiece of the mission; why not send it by itself? The stripped-down mission cost only a third of VOIR. In 1986, the new mission, initially called Venus Radar Mapper, was renamed Magellan after Ferdinand Magellan, the Portuguese explorer who attempted to circumnavigate the world in the early sixteenth century. Although the Magellan mission was a pale shadow of the original VOIR concept, the $270 million it required was still substantial, and proponents had to use a certain political dexterity to get it off the ground. In the end, Magellan proved to be a major triumph for the American space program and paved the way for future missions like it.

With their budgets squeezed, the engineers and scientists were forced to rethink their priorities and strategies. Fortunately for them, there were plenty of spare parts lying around from NASA's glory days that could be "recycled" in the new era of budget-conscious space exploration. For example, instead of having two antennae, *Magellan* had one, and it was a spare left from a Voyager spacecraft. *Magellan* would consequently have to flip regularly as it orbited Venus, acquiring data from the surface and then retransmitting it to Earth, but one antenna kept costs down. *Magellan* engineers also came up with ex-Voyager thrusters and a spare spacecraft body (not to mention surplus computers from the Galileo mission to Jupiter). Although the mission to Venus was beginning to look like a poor-man's deep-space extravaganza, the budgetary restrictions just seemed to raise the game for the engineers. Forcing them to think about costs brought an improvement in the sophistication and robustness of the software for the spacecraft, an altogether more subtle approach than resolving problems by throwing money at hardware.

Despite the mission engineers' skill and ingenuity, there was no way to guard against unforeseeable contingency. The Magellan master plan called for the spacecraft to be launched from the space shuttle *Atlantis*, the first deep-space probe ever to be so launched. It seemed the ultimate justification for the shuttle; after all, the shuttle was built to be a truck into space. But on the morning of January 28, 1986, NASA's dreams shattered when, seventy-three seconds after launch from Pad 39 at Cape Canaveral, Orbiter Vehicle 99—the space shuttle *Challenger*—exploded. A stunned world watched in disbelief as the apparently intact crew compartment separated from the fireball, fell out of the sky, and smashed two minutes later into the Atlantic Ocean.

NASA cancelled all space shuttle launches for almost three years while the accident was investigated. *Magellan* had been slated for launch in May 1988, and there was no way the mission would proceed on schedule. When space shuttle launches restarted in 1989, *Magellan* found itself in competition with the *Galileo* mission to Jupiter, which needed *Magellan*'s October 1989 shuttle launch window to satisfy its more elaborate deep-space trajectory. Because of the complexity of the flight path to Jupiter, the *Galileo* mission got priority, and *Magellan* was moved up to an earlier launch window in May 1989. Because of the intricacies of orbital mechanics, however, the earlier launch date for *Magellan* actually meant that it would arrive at Venus later than the mission scientists had planned. The longer transit time also carried with it the nagging worry that there would be more opportunities for things to go wrong.

*Magellan* was launched from *Atlantis* on May, 4, 1989, the first space probe ever to be launched from a space shuttle. It was America's long-awaited return to deep-space exploration, for it was the first interplanetary mission since the Pioneer Venus expedition over a decade earlier. The engineers had been right to worry about the long transit time; en route to Venus, *Magellan* developed overheating and orientation problems. The critical overheating issue was cured, though, by rotating the spacecraft and using its purloined antenna as a sunshade.

Despite this and other problems, *Magellan* successfully arrived in a polar Venus orbit in August 1990. The orientation of the orbit was critical, because *Magellan* would orbit north-to-south while Venus rotated east-to-west beneath. In this way, a complete, high-resolution map of the planet would be completed in just two Venusian days (243 Earth days), while allowing for plenty of redundancy.

The decision to spend just two Venusian days mapping the planet turned out to be a good one, since, after achieving its orbit, the mission was beset by technical problems. By the end of the first Venusian day, *Magellan* had only captured 84 percent of the planet. The engineers ingeniously fixed most of the problems, however, and by the end of the second Venusian day, 97 percent of Venus had been mapped.

*Magellan* was tasked with answering some specific questions. Most important among them involved the exact nature of the Venusian geology. Is Venus dominated by a plate tectonic regime similar to that of the Earth, or does it experience something different? Does erosion shape the surface of Venus in a similar manner as on Earth and (as was suspected at the time) on the surface of Mars? What is the surface of Venus made of? Another question harked back to the nagging uncertainties about Venus's incredibly hostile atmosphere: Had its greenhouse effect always existed, or had it developed over time? If the latter, could this greenhouse effect develop on other planets such as our own?

The results from the *Magellan* mission were nothing short of spectacular. An immediate result was that those who believed that Venus is a world without plate tectonics were vindicated. *Magellan* showed a world with no evidence of plate boundaries—the crucial areas where new crust is subducted back into the interior of a planet. Nor does Venus have any equivalent to the midocean ridges of Earth, where new material from the interior of our planet wells up, cools, and solidifies to form new crust. Instead, *Magellan* showed a world that lacked Earth geology's distinct partitioning of oceans, on the one hand, from continents, on the other. On Earth, the primary lithology of the oceans is basalt, whereas the primary lithology of the continents is granite. Venus, however, is almost exclusively basalt—as the Soviet Venera landers had demonstrated years before. Without plate tectonics, Venus is, functionally at least, only half the planet the Earth is.

Given the absence of plate tectonics on Venus, how does the planet's surface renew itself, providing that 500 million-year-young face? The answer is that, whereas on Earth volcanism works hand-in-hand with plate tectonics to shape the topography of the planet, on Venus volcanism is king and works alone.

*Magellan* showed that at least 85 percent of the surface of Venus was composed entirely of volcanic flows. The remainder, only 15 percent, is comprised of highly deformed mountain belts. Whereas the Earth has ocean trenches seven miles deep and mountain ranges five miles high, the total range of height from deepest to highest on Venus is within one kilometer of the mean radius of the planet. *Magellan* also found that Venus has no erosion. On Earth, erosion is driven primarily by water; on Venus, the lack of water means there is no erosion at all. On Venus, surface features routinely persist for hundreds of millions of years as opposed to only thousands on Earth. It now seems that Venus lost its water to space several billion years ago because of the planet's proximity to the sun and because of the runaway greenhouse effect. Without water, there were no carbonate minerals in the ocean (as on Earth) to soak up the carbon dioxide. The combination of these two factors plunged Venus into the spiraling greenhouse effect that today is responsible for its colossal temperatures. It dramatically shows how distance from the sun is a major parameter in making a world habitable—as we will see in the next chapter.

The Magellan mission was not the latest mission to Venus. The European Space Agency's Venus Express mission arrived in April 2006 with a comprehensive suite of instruments to further study the planet. *Venus Express* is currently investigating the reasons for the very high wind speeds in the upper atmosphere and looking at the processes that control the circulation of hostile gases in its atmosphere. *Venus Express* is also investigating whether there is currently volcanic activity on the planet.

The more we have studied Venus, the more we have realized that the superficial comparisons drawn in the era before exploration with spacecraft were wrong. The Venus experience proves that the only way to successfully understand other worlds—even those close to home—is to visit them.

# 4

# The Wizards of Earthsea: Earth and Moon

The science of comparative planetology is less than two decades old. With its development, the Earth has assumed an important role as the golden standard against which all other planets are measured. What we think of as normal on a planet—water, a breathable atmosphere, life—cannot be considered absolute normality. They are merely the local conditions we are used to; it is quite possible that the Earth is a most unique individual. Nonetheless, there are two ways the Earth provides a starting point for understanding our planetary neighbors: by providing a sense of time and, crucially, of the *scale* of time.

The immensity of geological time is one of our most important discoveries, and one of our most useful conventions is the device we use to slice geological time into units: the geological timescale. The Earth formed at the same time as the other planets, about 4.5 billion years ago. The universe, of which our solar system is but a part, is even older—approximately 13 billion years old.

It is convenient that all the planets of the solar system are the same age, because it allows scientists to use Earth's geological timescale to calibrate the timing and ordering of events on other planets. One of the greatest ironies in calculating these events is that the timescale we use is based on the one thing we have on Earth that no other planet has: life.

The divisions of Earth's geological timescale are named haphazardly, a consequence of the human and historical contingencies that shaped the science of geology and therefore the way the geological periods were determined. The earliest method of dividing geological time was devised by the Dane Nicholas Steno in 1669. He based his method on the way that fossils change systematically in rocks. William Smith, often described as the founder of modern geology, took the notion a step further and pointed out that rock types with assemblages of similar fossils could be found in widely separated areas. These rocks could be correlated, or identified as being of the same age. Biostratigraphy, as Smith's discipline became known, slices up geological time using the changing pattern of organisms in rocks.

There is more than one type of stratigraphy, however. Rocks can be divided on the basis of their composition and color as well, a discipline known as lithostratigraphy. In more recent times, geologists recognized that the direction of magnetism or the chemical composition of rocks can vary systematically, giving rise to the twin disciplines of magnetostratigraphy and chemostratigraphy. But these variations are all elaborations on biostratigraphy, the relative ordering of rock strata using fossils.

William Smith advanced the basic tenets of bio- and lithostratigraphy when he invented one of the most magnificent achievements of modern geology, the geological map. Using colors to identify areas with rocks of similar age, he mapped the distribution of rocks according to their ages, the jumping off point for understanding the geological history of any particular area.

Biostratigraphy is somewhat arbitrary because it assigns a general age, which is correlated to the rock's position relative to the subjective divisions of geological time. With the scientific advancements of the nineteenth and early twentieth centuries, a new method, geochronology, was developed that determines the absolute age of rocks. Geochronology uses radioactive decay as an absolute chronometer; the half-life of a radioactive isotope is known and the quantities of daughter products measured, giving the rock's absolute age.

At the present time, our understanding of the history of other planets—with the notable exceptions of our moon and Mars—is very primitive, but the combination of the geological map and stratigraphy, which have given us such a secure understanding of Earth's history, is now beginning to be applied to the other planets and moons of the solar system, allowing us to begin to unravel their history. In the not so very distant future, we will have a hybrid construct of historical events that will combine the histories of multiple planets, giving us a more complete picture than one planetary history alone can do. When we are able to add geochronology to that mix, which can only be done by analyzing material in situ, we will have begun to synthesize the history of the solar system. In the same way that William Smith's map revolutionized geology, this hybrid construct will revolutionize comparative planetology.

Our knowledge of the earliest events in Earth's history is obscured. Paradoxically, what we do know about Earth's early history comes from the study of our nearest neighbor, the moon. Earth's plate tectonics continually resurface the face of the planet, obscuring the history that would otherwise be written on the planet's surface and forcing scientists to look to the moon for answers. Without active plate tectonics to obscure the record, the moon's face shows events back to the earliest days of the solar system, including details about the Heavy Bombardment Period. Crater aging shows that, during this period, which ended about 4 billion years ago, bodies in our solar system were pummeled by a colossal quantity of meteorites and asteroids.

At the end of the Heavy Bombardment Period, life appeared on Earth. The controversial evidence for this comes from the rocks of Akilia Island on Greenland's west coast. Some scientists claim that these rocks—part of the Isua geological complex—show when life began not on the basis of conventional fossils such as bacteria but on the presence of graphite particles in the fabric of the rock. They claim that these particles show a very negative carbon signature, which marks the presence of photosynthesis and, therefore, life.

Photosynthesis is the synthesis of sugars from carbon dioxide using sunlight. It is one of the oldest chemical tricks in life's cookbook and has the happy consequence of liberating oxygen, a requirement for the evolution of complex life. Photosynthesis preferentially uses the light form of carbon, carbon–12, to make new organic material, which leaves a characteristic signature in rocks suggesting that photosynthesis was present. The carbon signature is an important tool in the search for life on other planets.

If the Akilia Island claims are correct, the implications are astounding, because they suggest that life began on Earth almost as soon as the heavy bombardment of the inner solar system ended. In other words, life began on our planet as soon as conditions that could support it prevailed. If true, it suggests that life is not rare, as it has often been considered, but, at least on this planet, inevitable. All of which rather begs the question—what is life?

## EDDIES IN THE ENTROPY GRADIENT

Life is surprisingly hard to define, and some of the criteria cited must be used in conjunction with others. For example, one of life's characteristics is that it replicates—it makes copies of itself. This criterion alone cannot define life, however, because some other substances—copper sulphate and other inorganic compounds that crystallize—do the same thing. Another oft-cited criterion for life is the possession of a metabolism, or the tendency of living creatures to extract energy from their environment to fuel life's processes. On Earth, we recognize two major categories of metabolism: autotrophs, which extract energy from sunlight or chemicals, and heterotrophs, which obtain energy through the consumption of autotrophs. But once again, this, in and of itself, is not a sufficient definition—just think of fire, an inorganic chemical reaction that extracts energy from the environment.

A definition of life that is tenable, however, involves the Second Law of Thermodynamics. One of life's characteristics is that it builds order from disorder. Babies are constructed from eggs and sperm. The fossil record shows that animal dynasties generally increased in complexity from the days of the early Cambrian to the present day, even if evolution, as we have seen, did not proceed in straight lines. And, just because humans are the current model of choice in the advanced life form stakes, there is nothing to suggest that evolution is not currently developing more and more complex beings. This trend stands in opposition to the Second Law of Thermodynamics, which states that the tendency of all ordered systems is to move from a state of complexity to one of simplicity. This tendency is called entropy. Life is a local reversal in the entropy gradient—a pocket of winding-up in a universe that is running-down.

Some, such as Richard Dawkins, put more weight on another feature of life that is uniquely characteristic: its susceptibility to evolution by natural selection as formulated by Darwin. Life produces variation, individual offspring variably adapted to success in the environment. Those who are better adapted—who are most fit—survive and pass on their successful genetic code to their descendents. In this way, favorable genetic combinations accrue over time. There is nothing magical about natural selection; it is a simple winnowing of variation through competition. It is this process, however, that generates the local reversal of the entropy gradient. In other words, life is the normal processes of physics and chemistry filtered through the refracting lens of Darwinian natural selection.

The related negative entropy/natural selection definition of life applies not only to planet Earth but applies across the universe. These general principles underpin the phenomenon we know of as life.

It is appropriate to ask at this point whether it is "merely" the possession of life that makes Earth unique among the other planets of the solar system. The answer is no. As we shall see, there may have been life on Mars in the past, and there may yet be life on moons such as Jupiter's Europa. In fact, Earth is no more or less unique than any other planet of the solar system—it is merely itself. The fact that Earth has liquid water in abundance is almost certainly the reason life developed on our planet, but other planets and moons have—or had—liquid water, too (Europa and Mars once again, for example). At least Mercury has a magnetic field like the Earth, and Mars probably once had plate tectonics. Venus certainly has volcanism. The truth is that, given a biographer's eye—that is, a sense of history—our Earth is not unique. It is merely where the action is at the moment.

## THE TIME TUNNEL

No one is quite sure how life got started on our planet, but we can make an educated guess. At some point about 3.8 billion years ago, molecules capable of making copies of themselves formed from the prebiotic building blocks present in the early atmosphere. We know this much from the experiments of Stanley Miller and Harold Urey in the 1950s, who showed that the organic precursors of life—amino acids—could be formed in a simple laboratory apparatus set up to mimic early conditions on Earth. These conditions consisted of a strongly reducing atmosphere—one without oxygen—with a continuous exposure to electrical energy in the form of lightning. This atmosphere was composed of carbon dioxide, carbon monoxide, methane, ammonia, and other equally unpleasant compounds, which illustrates my earlier point that what we are accustomed to is not particularly unique to this planet, it is just unique to this planet *at this stage* in its history.

If we scale the duration of Earth's history against one day, with midnight being the formation of the Earth 4.5 billion years ago and midnight twenty-four hours later being the present moment in time—our today—we get some conception of the immense span of geological time. On this scale, life started at about 3:12 AM, definite bacterial fossils appeared by 7:28 AM, the Cambrian explosion occurred at 9:07 PM, the dinosaurs became extinct at 11:40 PM, and humans appeared at 11:59 PM—one minute before midnight! The huge time gap between the appearance of bacterial fossils and the Cambrian explosion underlines something that all paleontologists worth their salt will tell you: this planet belongs to bacteria, and it always has.

Any attempt to conceive what the first life forms were like, however, becomes a "which came first, the chicken or the egg?" conundrum. We are hidebound by what we know today about the structure of life, whereby DNA is required to synthesize proteins, and protein catalysts are required to synthesize DNA. One cannot operate without the other, yet to imagine the simultaneous emergence of two such complex and interconnected systems on the prebiotic Earth stretches and then shatters credulity. A way around this impasse is to propose that there was a set of molecules that came before DNA that did not require the presence of a protein catalyst for its synthesis. It is known that RNA on its own is capable of self-replicating and that it may have had a precursor called TNA (threofuranosyl nucleic acid). Both of these molecule types can replicate without proteins, and TNA is much simpler in structure than RNA. But whatever their structure, these self-replicating molecules were alive in the Darwinian sense in that they could make accurate (but not perfect!) copies of themselves and evolve by natural selection. Once that process had started, life was truly underway.

These first "organisms" were not what we would recognize as alive in any conventional sense. They were self-replicating molecules in a complex biological soup. Eventually, these naked, self-replicating molecules became enclosed in a vitally important membrane; a series of lipid molecules arranged themselves with the water-loving and water-hating ends aligned together, perhaps organized by something as simple as the tendency of various types of clay to hold molecules in particular orientations. This development may have been the origin of the cell membrane, one of the most important innovations in the evolution of the cell. Self-replicating molecules now had a structure to protect themselves from the environment and to aid their internal biochemistry. They were the first prokaryotic bacteria—bacteria that have no nucleus and no true cell wall. Do we have any evidence for their existence? That question is one of the most controversial in the earth sciences.

Far in the northwest of Australia is a landscape that looks like the setting for the Mad Max movies. It is a land of choking dust and razor-sharp spinifex grass. The long trek from the town of Marble Bar out to the eroded, desiccated streambed known as Chinaman Creek makes you feel you are already on Mars. It is a land of spectacular desolation. On the approach to Chinaman Creek, there are outcrops of glassy-looking rock on the edges of the rills and gullies. Here you will find the Apex Chert, some of the most ancient sedimentary rocks in the world.

Back in the 1980s, a young scientist from the University of California at Los Angeles came to this blighted place in search of fossils. Bill Schopf was interested in the history of Precambrian life, that long era of Earth history spanning the Achaean and Proterozoic—each of which dwarfs the duration of the Phanerozoic that started some 550 million years ago with the rise of conventional fossils at about the time of the Cambrian explosion. Despite the derision of colleagues who believed he would never find bacterial fossils in such a place, Schopf was a seasoned veteran of campaigns to separate bacterial fossils from the clutches of deep time. He'd already been involved in finding convincing bacterial fossils in North America's 2 billion-year-old Gunflint Chert. If such early Proterozoic bacterial fossils could be found, why not even older ones?

Schopf knew there were only two places in the world where potentially fossiliferous rocks of this very early age might be found: in the Western Australian Pilbara Supergroup, or in the hills and valleys of South Africa's Barbeton Mountain Land, the Swaziland Supergroup. Both of these sedimentary sequences have been somewhat altered by post-depositional heating and sinking within the Earth's crust, but the Pilbara Supergroup was less affected than the Barbeton Range; hence, Schopf decided on Australia.

The outcome of his studies was the publication of a 1993 *Science* article detailing eleven microfossils that he claimed were the oldest fossils in the world. They were all thought to be members of the prokaryotic cyanobacteria (blue-green algae) and, at 3.5 billion years, were only 300 million years younger than the Isua carbon fossil. If Schopf was correct, they must come from the age soon after the cell membrane was acquired. In the last few years, however, a vitriolic disagreement has erupted over the interpretation of Schopf's fossils. A group based at Oxford has suggested that the fossils cannot be bacteria because the structures show branching, a characteristic no other bacterium is known to possess. Like the controversial carbon isotope signatures of the Akilia enclave, the debate on Schopf's findings has not been settled, but both cases highlight the difficulties of unequivocally identifying life at these extremities of space and time. It brings into question whether we will ever be successful in our search for life (fossil or extant) on other planets if we have this level of difficulty on our own.

So far we have been concerned with prokaryotic bacteria—bacteria without a true nucleus and that lack internal organelles. At some point in the long period of time known as the Precambrian, when the highest form of life on our planet was primitive pond scum, the third great innovation in life's history occurred—the evolution of the eukaryotic cell.

My colleague Linda Partridge has described the eukaryotic cell as "a committee" of prokaryotic cells. This is a perfect description because the eukaryotic cell is larger than the prokaryotic and has several substructures. Eukaryotic cells—such as mitochondria and, in some cases, chloroplasts—were probably themselves once free-living prokaryotes that one day came to appreciate the advantages of communal living. This concept was coined by Lynn Margulis, Carl Sagan's first wife and a formidable intellectual. She named this theory the endosymbiosis theory of eukaryotic origins, and part of its fascination is that its basis lies as much in the search for life on other planets (on which Sagan made his reputation) as in any aspect of conventional biology. The endosymbiosis theory is one of the first great triumphs of the new science of astrobiology.

Mitochondria handle the vitally important energy budget of a cell, and chloroplasts handle photosynthesis—the ability to make sugar out of sunlight and carbon dioxide. Arguably, the evolution of the chloroplast via endosymbiosis was what paved the way for the biggest change in the nature of the Earth: the development of an oxygen atmosphere. For more than thirty years, scientists have believed that atmospheric oxygen first reached appreciable levels between 2 and 2.4 billion years ago. Geologists have labeled this founding moment in Earth history the "Great Oxidation Event" (GOE). In rocks older than 2.4 billion years, an unusual chemical signature in the rocks indicates a low-oxygen atmosphere. But questions remain, most significantly, why did atmospheric oxygen climb to significant levels at the 2.4 billion-year marker when oxygen-producing bacteria had evolved by at least 2.7 billion years? A second, equally puzzling question concerns why the ratio of carbon isotopes in marine carbonates did not rise as the GOE kicked in. Today, isotopically light carbon (carbon–12) is removed from the ocean-atmosphere system by photosynthesis and then buried as organic carbon. The carbon isotope record of marine carbonates is thus heavier than it would be in the absence of photosynthesis. In principle, we should be able to use this proxy to definitively determine the GOE's onset. Yet, apart from a brief period of carbon–13-enriched rocks following the GOE, carbonate rocks that formed before and after 2.4 billion years ago show broadly the same isotopic signature.

According to the most recent theory, once photosynthesis began, the atmosphere could exist in only one of two end-member states: in a low-oxygen or high-oxygen state. This bistability resulted not from variations in the rate at which oxygen was being produced but, rather, by variations in the rate it was consumed. Scientists believe that, at the start of the GOE, the newly minted oxygen was rapidly consumed in ultraviolet-catalyzed reactions with the methane produced by living organisms. But as oxygen levels increased, so, too, did the level of ozone in the atmosphere, which gradually shielded the surface and the lower atmosphere from this solar ultraviolet radiation. As this radiation was reduced, oxygen consumption decreased, and oxygen began to accumulate in the atmosphere. In the meantime, however, the atmosphere was stuck in a low-oxygen state for a long time following the evolution of photosynthesis and was held there until the ultraviolet radiation was deflected by the accumulation of ozone.

An additional factor leading to the buildup of atmospheric oxygen may have been a change in the control levied on it by the burial of organic carbon in the oceans. Oxidation of deep ocean sediments is a major consumer of oxygen; thus carbon burial exerts a ferocious hold on atmospheric oxygen levels. With a slight decrease in the oxidation rate of carbon in the deep ocean—perhaps caused by the deposition of unoxidized carbon-rich sediments on the continental shelves—marine and therefore atmospheric oxygen levels would increase very quickly. The lack of evidence indicating such a decrease in the carbon isotope record is most likely due to our inability—even with the most sophisticated techniques—to detect such a small change (3 percent) over such a colossal amount of time. Regardless, once started, the momentum of this chemical rollercoaster was irreversible, and the Earth was squarely set on the road to oxygenation. Possibly, after the evolution of photosynthesis, oxygen levels fluctuated on a small scale before the ozone and carbon burial brakes were finally removed. One thing, however, is clear: the evolution of Earth's early atmosphere seems much more complicated and delicate than we once thought. It was not simply a matter of lighting the photosynthetic fuse; far more subtle chemical and biological processes had their parts to play, too.

For the next 2.5 billion years, Earth became accustomed to having a blanket of oxygen around it. Strange features were added in that era. Great layers of banded iron (Banded Iron Formations or BIFs) were periodically laid down in various parts of the world's oceans. These BIFs formed as the vast amounts of iron in Earth's early acidic oceans precipitated out as insoluble iron oxides as the oxygen concentration grew. As the deep oceans approached oxygen saturation, small variations in surface water oxygen produced local, shortlived alterations in marine and probably atmospheric oxygen concentrations, which explains why many of these BIFs alternate with bands of iron oxide (laid down when oxygen concentrations were high) and layers of iron-poor shale and chert (laid down when oxygen concentrations were poor). Today, these great stores of iron provide the raw material for the steel used in the world's economy, providing testament to the long-ago efficiency of a Gaian chemistry in transition.

This long period of the Precambrian is known as the Proterozoic, which falls after the Hadean and Archaean and before the Phanerozoic, the era of visible life. Australia has some of the best Proterozoic sequences in the world. We have studied and analyzed them closely, but mostly what can be said of this period of Earth history is that nothing much happened. The stage was being set for what was to come.

The Proterozoic offers an important lesson for our future studies of other planets' histories. When we finally go to other planets, it is entirely likely we will find worlds where nothing much seems to be happening. The message for our astronaut descendents is not to judge too hastily but to evaluate with Earth's history in mind. Like an army on the eve of battle, long periods of inactivity are punctuated by frantic periods of shortlived activity. Blink and you might miss it.

But something did happen toward the end of what we call the late Neoproterozoic (about 600 million years ago), which, like the rocks of Pilbara, is an Australian story, too. Shortly after World War II, a young government geologist named Reg Sprigg was prospecting for the new mineral of choice during the nuclear age: uranium. In the Ediacara Hills of South Australia, Sprigg found fossil impressions in sandstone that appeared to resemble, of all things, a patchwork quilt. Today they are known as the Ediacaran fauna, and we believe the organisms were probably related to jellyfish.

Jellyfish make up the only animal group on Earth that is diploblastic: they are composed of only two layers of body tissue. All other groups are triploblastic, that is, composed of three layers of body tissue. Ediacarans are one of those fascinating waystations in life's history. They might even have been top dog if evolution hadn't discovered that three layers of body tissue impart more evolutionary advantages. After all, with three layers you can make organ systems, not just tissues. You cannot fool natural selection.

The three-layered animals that flowered immediately after the jellyfish were much more sophisticated and, consequently, much more successful. These were the animals of the Cambrian explosion, a short period of time, probably less than 10 million years in duration, during which all the animal types we are familiar with today evolved in a single big bang. No one really understands the events that stimulated this explosion, the most momentous evolutionary event since the development of photosynthesis. It may have had to do with several global ice ages. When I say global, I mean global. From various lines of evidence that include, once again, the subtle distribution of carbon isotopes as well as lines of fossil magnetism preserved in ancient rocks, it is quite clear that the Earth was covered in ice—or at least slush—as far south as the equator. Proponents of the full-blown version of this theory call it the "snowball Earth." Others, who accept the lesser version of the theory, call it the "slushball Earth." If either of these theories is even close to the truth—and there is compelling evidence that they are—there can be no doubt that there was terrific evolutionary pressure on organisms at that time. The version of evolutionary theory known as punctuated equilibrium predicts that populations experiencing restricted gene flow will undergo bursts of incredibly rapid evolution. Accordingly, at the end of the snowball (or slushball) period, life on Earth exploded into action—the Cambrian explosion.

This is unlikely to be the whole story, because the animals of the Cambrian explosion did not spring out of nothing. Evidence in the topmost rocks of the latest Proterozoic shows that quite complicated, if tiny, multicellular animals inhabited the seas of that time. The Cambrian explosion may have been their way of appearing more prominently on the world stage by simply increasing their size. Again, this is a necessary oversimplification. The techniques of molecular biochemistry allow us to compare animals and determine their differences. By comparing genetic differences, we can make inferences about when major groups separated from each other in the fossil record.

The most famous Cambrian explosion fossil locality in the world is the Burgess Shale of British Columbia. These rocks represent a snapshot in time only twenty or so million years after the Cambrian explosion, offering insight into what probably happened during the explosion itself. According to a hypothesis made world-famous by Stephen Jay Gould in his elegiac book *Wonderful Life*, the animals of the Burgess Shale may have been much more diverse than present-day animals. While hotly contested, Gould's theory seems particularly significant, especially when put together with recent genetic findings. These studies suggest that, about the time of the Cambrian explosion, a major group of genes, called Hox genes, common to all triploblasts underwent a radical change that allowed for the diversification of a wide range of body blueprints. The profound proliferation of new animals during the Cambrian explosion may thus be a direct result of the increase in the amount of raw materials evolution had to select from, in addition to the extreme environmental variability.

Although no one would say this too loudly to a paleontologist, after the seismic evolutionary upheavals of the Precambrian—from the evolution of replicating molecules at its start to the development of triploblastic animals just before its end—events during our current era, the Phanerozoic, are much more ho-hum. It is as if evolution made all the big decisions about the form life would take on our planet before the era of "visible" life started.

The early Paleozoic saw the gradual colonization of the seas by complex invertebrate animals. By the middle Paleozoic, the land was beginning to be colonized by terrestrial green plants descended from photosynthetic marine algae. Toward the end of that period, the land was invaded by vertebrate animals that also took the opportunity to exploit these new ecological niches.

The Permo-Triassic boundary was a major extinction event—indeed, the most significant such event in the fossil record—that marks the end of the Paleozoic era and the beginning of the Mesozoic era. This event allowed for the diversification of a group of terrestrial animals that would become the mammal-like reptiles. For an unbelievably long period of time, their mammal descendants were spectacularly unsuccessful amid a world dominated by reptiles and dinosaurs. As discussed in Chapter 6, however, the dinosaurs' domination ended when a rogue asteroid smashed into the Earth 65 million years ago and reset evolution's clock.

In the aftermath of that cataclysmic asteroid encounter, which marks the end of the Mesozoic period and the start of the Cenozoic, mammals took center stage for 60 million years. At the end of that time, the onset of another—albeit lesser—series of ice ages provided another stressful environment. These extreme climate changes favored large brain size in apes. Fast forward to the present day, and those apes' descendants have invented fire, the wheel, language, music, nuclear energy, computers, and machines to explore the solar system.

## SIZE, SHAPE, AND STEPHEN JAY GOULD

In 1982, as a university sophomore, I encountered a book that changed my life: *Ever Since Darwin*, the first volume of Stephen Jay Gould's eloquent, entertaining, and groundbreaking essays on evolution. They should be required reading in every high school.

One of the things I most admire about the early writings of Stephen Jay Gould is that, in those days, he thought the universe could be explained by relatively simple physical laws. Gould's earliest work was on the relationship between size and shape in animals and the way the two are related. One of his most famous essays is on the antlers of the Irish elk—*Megaloceros giganteus*—in which he shows mathematically that the antlers are as big as they could physically get and still allow the male elk to carry their weight. His point is that there are physical laws to biology, and they cannot be violated.

Steve Gould could not resist extending his fascination with scaling laws onto a bigger canvas, and he devoted one of the essays in *Ever Since Darwin* to the relationship between size and shape of planets. In "Planetary Sizes and Surfaces," Gould's preoccupation is with the fact that the area of surfaces scale to the second power (that is, for example, 2 × 2), whereas volumes scale to the third (2 × 2 × 2). A small planet or moon will thus have a relatively higher surface area to volume ratio than a larger planet. Gould was convinced that this simple geometrical expression held the key to explaining the nature of planetary surfaces, and he used his theory to explain why the surfaces of small bodies like the moon and Mercury are heavily cratered while the Earth's face is relatively smooth and Mars, of intermediate size, has both cratered and smooth regions. Gould argued that the density of cratering was intimately linked to the size of the planet, because the planet's size controls the magnitude of the extrinsic and intrinsic forces that shape surface structures. The Earth is thus large enough to have an appreciable gravity field that, in turn, holds gases and prevents them from being vented into space. These gases are responsible for the erosion on our planet's surface by wind and by water. The Earth is also large enough to have retained an active core of radioactively decaying materials. This energy source powers the upward convection of molten rock, which ultimately drives the momentum for plate tectonics. These extrinsic and intrinsic forces are consequently responsible for the lack of large craters on the Earth. Erosional processes gradually smooth the scars of impacts, while plate tectonics endlessly recycles the ocean basins and continents back into the mantle. The Chixchilub impact structure—the scar from the asteroid that wiped out the dinosaurs—exemplifies this action; it is currently being subducted under the Yucatan Peninsula by plate tectonics.

Both the moon and Mercury are small planetary bodies. The moon has never had an atmosphere, and Mercury's leached away early in its history. Consequently, neither body has experienced the erosion needed to clean away the scars of meteorite impacts. Furthermore, both the moon and Mercury are tectonically quiescent. Moonquakes detected by the instruments left behind by the Apollo landings indicate that they occur almost a thousand kilometers deep—ten times deeper than Earth's own internal tremors. Being small, the internal heat of the moon and Mercury ran out eons ago; hence no tectonics cleanse the surface.

Mars is intermediate in size, and broadly half its surface is cratered while the rest is relatively smooth. This face reflects the activity of rather limited intrinsic and extrinsic resurfacing engines. Being smaller than the Earth, Mars's gravity is only strong enough to hold an atmosphere about two hundred times thinner than ours. Yet even this thin atmosphere can be an agent of significant erosion when it is whipped up into long-duration dust storms. As Carl Sagan pointed out, photographic evidence from probes has shown compelling evidence for the presence of water; the pictures show erosional features that look identical to those produced by the action of liquid water on our own planet. Sagan's observation—based largely on *Mariner* data—was verified in the last couple of years with the images from *Mars Express.* These pictures show incontrovertibly that there was once water—and therefore fluvial erosion—on Mars.

Evidence for the action of intrinsic factors on Mars is also plentiful. Mars has a volcanic province with giant mountains bigger than anything on Earth. Olympus Mons is 500 kilometers wide at its base, 8 kilometers high, and has a volcanic caldera 70 kilometers in diameter. The nearby Vallis Marineris gorge dwarfs anything on Earth. It is 120 kilometers wide, 6 kilometers deep, and more than 5,000 kilometers long. How else to explain these features except by concluding that Mars once had some form of plate tectonics?

Gould believed that Olympus Mons is a static Martian analogue of the Hawaiian Islands chain. The Hawaiian Islands stretch from the northwest to the southeast across the North Pacific ocean, with the youngest and therefore most active volcanoes at the southeastern end. This chain effect is due to the movement of the Pacific plate, which for millions of years has been moving above the Hawaiian superplume. Like a candle underneath a waxed plate, the volcanoes are the surface scars of this subterranean movement. If crustal movement on Mars is slow because of its intermediate size—as Gould believed—then any volcano formed in this fashion would just pile itself into a supervolcano like Olympus Mons. Gould also argued the Vallis Marineris was an abortive attempt at tectonic rifting.

Gould wrote this essay early in his career. It was a first attempt to relate the sizes of planets to their natures, and therefore it was one of the most daring and important contributions to early planetary geology. Even more extraordinary is that it was formulated by a man who was fundamentally a biologist and geologist rather than a comparative planetologist.

Gould was able to see how well his pet theory fared during his own lifetime as the Voyager missions to the outer planets expanded our understanding of worlds beyond our own. He eventually accepted that it was an oversimplification, but there is no doubt that it was a useful hypothesis at the time. But lest we get ahead of ourselves, we must first tend to unfinished business closer to home—our own moon, Mars, and the asteroid belt.

## EARTHRISE

Not so many years ago, I stood with my family at the base of a gaunt steel structure baking in the heat of a Floridian summer. Launch Pad 34-A at Cape Canaveral Air Force Station has long been decommissioned, but there is an indefinable melancholy about the place. Before approaching the site, the NASA tour guide asked gentlemen to remove their hats. And so we stood bareheaded beneath the skeletal steel ruins that jutted monstrously toward scudding clouds.

On January 27, 1961, this place saw the first tragedy of the U.S. space program when Guss Grissom, Ed White, and Roger Chaffee lost their lives in the *Apollo 1* fire. In those days, the Apollo capsules were highly primitive prototypes, and faulty wiring sparked and ignited the interior of the capsule that was pressurized with pure oxygen. The capsules at that time had no quick release mechanisms. The three astronauts' fates were sealed.

Pad 34-A still stands today as a mute testament to the heroism of the Apollo era. It is a haunting place, but it is part of the story of lunar exploration—a story as old as the telescope.

By November of 1609, Galileo had started one of his earliest projects with his new telescope, a detailed map of the near side of the moon. Galileo's observations immediately showed that the moon's surface was heavily cratered with mountains and valleys. Dark areas of the moon clearly visible to the naked eye are called "Maria." Famous examples are Mare Tranquillitatus (the Sea of Tranquillity) and Oceanus Procellarum (the Ocean of Storms). These dark areas are now known to be vast tracts of lowlands, and the lighter areas that surround them are the lunar highlands. The moon is one of the most heavily cratered objects in the solar system and, like Mercury, is tectonically dead. These craters number in the tens of thousands and range in diameter from 180 miles to the size of a pinhead. The larger craters have high walls and depressed, flattened floors; many have central peaks or ridges. From some, streaks of lighter-colored material radiate out (like the giant Copernicus crater in the moon's southern hemisphere). The moon also has domes, ridges, and rilles—long, sinuous channels that can run for several kilometers.

Visionary Nobel Laureate and Manhattan Project veteran Harold Urey stimulated modern scientific interest in the moon. In 1958 he approached the NASA associate administrator, Homer Newell, and suggested that the scientific study of the moon be a major focus of the newly minted American space effort. This conversation occurred three years before John F. Kennedy committed America to put a man on the moon by 1970. Crucially, Urey's interest was scientific, not political.

Urey believed that the moon was a primordial fossil left over from when the Earth was formed and that its craters resulted solely from meteorite bombardment (the so-called "cold moon" theory). This theory was in marked contrast to a competing one suggesting that the moon, like its neighbor Earth, had once suffered from intense volcanism and had been tossed off the primordial Earth while it was still forming (the "hot moon" theory). These "hot-mooners" believed that intense volcanism eons ago had turned the moon into an overcooked cinder, forming the moon's craters in the process, and had then ceased.

To investigate these ideas, and to start the search for suitable Apollo landing sites, NASA and the Jet Propulsion Laboratory launched a series of moon probes. The first of these probes were the Ranger series:

unmanned drones whose only function was to photograph the lunar surface before crashing into it. This series was followed by the Surveyor probes that soft-landed on the moon and took samples. In part, the Surveyors were investigating the nature of the moon's surface to ensure that it was sufficient to support the weight of the planned Apollo lunar modules. The geochemist Thomas Gold had warned—much to the derision of his colleagues—that the surface of the moon might consist of fine dust meters deep that could swallow a spacecraft whole. The final series of probes to precede Apollo flights were the Lunar Orbiters, tasked with taking high-resolution photographs of potential landing sites. They covered 95 percent of the moon's surface from a height of only twenty-eight miles, automatically developing and then scanning the seventy-mm film for transmission to Earth. They could resolve objects down to a foot in diameter.

The Apollo missions themselves were a spectacular success. The later missions especially (the so-called J-missions—*15*, *16*, and *17*) concentrated on scientific questions of lunar origins. The final mission, *Apollo 17*, included a fully qualified geologist, Harrison "Jack" Schmidt. It was the only mission to carry a scientist, and it was the final vindication of Harold Urey's contention that the moon was worthy of serious scientific study.

In the 1970s and 1980s, as a direct result of studying lunar samples brought back from the moon by Apollo missions, a third theory of the moon's origins was worked out by George Wetherill at the Carnegie Institution. Wetherill adapted calculations on the orbits of the asteroids to investigate how a group of small, randomly colliding rocky bodies orbiting the early sun might have behaved. Using newly developed multivariate statistical techniques, made possible only by the new generation of digital computers, he showed that hundreds of bodies with roughly the same mass as the moon could form one or two Earth-sized planets (that is, Venus and Earth) and perhaps a few smaller rocky planets (Mercury and Mars), orbiting at their observed distances from the sun.

In the mid-1970s these ideas led two groups of scientists, William Hartmann and Don Davis, and A. G. W. Cameron and William Ward, to propose that the moon's formation almost certainly involved one or two planets the size of Mars hitting the primordial Earth while it was still forming from the material of the accretionary disk. If such a giant impact occurred off-center, it would fire molten mantle material into an orbit around the Earth. They hypothesized that this debris later coalesced into the moon. It was an elegant solution to an old problem and is widely accepted today.

George Wetherill retired in the early 1990s but continued to work, enticed this time by the idea of planets around other stars, a concept then only on the verge of being proved. He showed that it was possible for Earth-like planets to form, even if they were much larger than the Earth. It was a crucial idea because it implied that worlds of high mass—which would be detected only a few short years after Wetherill's predictions—could form around other stars. Details remain to be explored, but this could yet be Wetherill's greatest legacy: a universal theory of the formation of habitable planets like our own.

It is currently fashionable to pretend that the crewed lunar landings of the Apollo era were nothing more than political stunts engineered by strutting Cold War warriors on a high-octane power-trip. In fact, nothing could be further from the truth. Among other monumental discoveries, the Apollo program proved that both the "hot moon" and the "cold moon" theories are correct. The moon, like our own planet, has a long history of volcanism, but it ceased millions of years ago. The meteorite impacts that occurred afterward are what formed the moon's characteristically scarred surface. Only on the moon can we see the first billion years of our solar system's history—a history common to all the planets. The later Apollo missions also discovered that the youngest moon rocks are as old as the oldest Earth rocks; that all moon rocks originated from high-temperature processes that did not involve water; and that early in its history the moon was melted to great depths, forming a planetary ocean of molten magma.

The J-missions also ventured into the rugged territories of the lunar highlands, including the Hadley-Apennine Mountains (*Apollo 15*) and the Descartes Highlands (*Apollo 16*). Interestingly, today there is a resurgence of interest in the science of the Apollo era. This is partly due to the recent imaging of the *Apollo 15* landing site, which has been positively identified as a discolored area by a moon-orbiting reconnaissance satellite known as the *Clementine Lunar Orbiter*. Those who claim that the Apollo missions were all hoaxes are currently enjoying a large, indigestible, and well-deserved meal of humble pie.

Let me say something in passing about moon landing denials, although I shall not dignify these assertions with more than a paragraph. Suffice it to say that to me those who promote such mischievous views are the empty-minded belittlers of one of the greatest scientific and technological achievements of the twentieth century. They demonstrate that ignorance is not only currently tolerated but encouraged by blurring the distinction between critical analysis and blind criticism (witness that the teaching of evolutionary theory is prohibited in many schools in the United States). To deny that the moon landings occurred is not only an insult to many brave and hard-working men and women but also destroys the joy and wonder of a major human achievement, an accomplishment that testifies that we did, however briefly, transcend our primitive ape minds and achieve something truly noble.

And so, as we leave the moon and head toward the deeper fastnesses of the solar system, let us not forget Pad 34-A. The sacrifice of Grissom, Chaffee, and White was not only political; it was scientific. Without the Apollo pioneers we would not have a theory of how our moon formed, and we would not have Wetherill's legacy—a universal theory of solar system formation. Next time you visit the Cape, try to see Pad 34-A. And don't forget to remove your hat.

# 5

# The Martian Chronicles: Mars

Of all the planets we have set out to explore in the past fifty years, there is only one—Mars—that we have a detailed picture of, and this picture has only emerged in the last decade. Mars has been the most intensively studied planet because we think of it as the one most like our own. It exercises our imaginations; we never tire of looking at it or thinking about it. We covet it.

If Venus is the Russian planet, then Mars is, without doubt, the American one. Early in the space age, the United States focused on Mars with an obsessiveness that has never subsided. More American probes have been sent to Mars than almost anywhere else in the solar system (with the exception of the moon). The moon landings, with the timeless images of American astronauts planting the Stars and Stripes on the lunar regolith, also provide a vital clue to the American fascination with Mars; the pioneer spirit continues to drive

Americans, and the frontier has been pushed beyond the bounds of Earth to the red planet, hanging low and baleful in the sky.

By the early 1970s, NASA had sufficient confidence from the early Mars Mariner missions to mount a head-on assault on the red planet with the Viking spacecraft. The Viking missions would not be a timid, orbital-only effort like *Mariner 2* or even like the cross-your-fingers-and-hope-for-the-best landing effort the Soviets had attempted with their *Mars 3* and *6* landers, both of which failed. Viking was nothing less than the planned soft-landing of two fully equipped, state-of-the-art, no-expenses-spared, chemical laboratories on the Martian surface. In the ultimate symbolic gesture of America staking her claim on the new Wild West, *Viking 1* and *Viking 2* were scheduled to land on the bicentenary of America's founding, July 4, 1976.

Ironically, America's love affair with Mars dates back sixty years to one of her most famous and controversial expatriates, Werner von Braun. During the last days of World War II, von Braun, who had evacuated with his team from the German rocket facility at Peenemunde to Oberjoch near the Austrian border where he surrendered to the American army, was transported to the White Sands Missile Range in the New Mexican desert to work on repatriated V2 rockets. In many ways, to von Braun and his associates the 1940s were "the desert years." During von Braun's stay in the relative wasteland of the Chihuahuan Desert, he wrote an essay titled "The Mars Project." It was a remarkably prescient document outlining a human mission to Mars that would require a flotilla of ten spacecraft crewed by seventy men. The spacecraft would be assembled in low Earth orbit from components ferried up from the surface. Once in Mars orbit, three of the spaceships would be equipped with wings. One of the winged ships would also be outfitted with skis, and it would descend to the Martian surface to land on the smooth frost of the polar regions. The crew would then drive to the equatorial regions to construct a landing strip for the other ships. Von Braun estimated that his plan would enable 149 metric tons of men and material to be ferried to Mars's surface within his lifetime. The piece, eventually published in the influential magazine *Collier's* in March 1952, provided an ambitious blueprint for American deep-space exploration. During the 1960s, von Braun became engrossed in the development of the giant Saturn V rocket, and plans for a Mars trip became secondary to the visits to the moon. But von Braun's ideas were not forgotten.

As we have seen, NASA was founded on July 29, 1958, with the avowed intent of developing spacecraft and high-altitude, high-speed airplanes. The new agency combined the functions of the World War I–era National Advisory Committee on Aeronautics (NACA) and the two new technologies of ballistic missiles and rockets. From the start, NASA's emphasis was squarely and unrepentantly on developing the technology needed to compete with the Soviet Union for control of the high frontier.

The first director of NASA, T. Keith Glennan, was an energetic and politically well-connected individual with a ferocious sense of purpose. Glennan quickly gained control of NASA and incorporated part of the navy's research laboratory near Washington, DC (creating the Goddard Space Flight Center) and the army's ballistic missile center in Huntsville, Alabama (which became the Marshall Space Flight Center). Also absorbed was Werner von Braun and his V2 rockets. The eight thousand employees of NACA had been spread mainly through three major research laboratories: the Langley Aeronautical Laboratory in the tidal flats of Langley, Virginia, the Ames Aeronautical Laboratory in California, and the Lewis Flight Propulsion Laboratory in Ohio. By the dawn of the 1960s, with these and other acquisitions from the many aerospace projects that littered the industrial landscape of postwar America, Glennan forged the mightiest aerospace organization the world has ever known.

Glennan gained control of one of his most significant acquisitions, the Jet Propulsion Laboratory (JPL) in California, quite early, in December 1958. The alliance of NASA and JPL became one of the monumental partnerships in the history of planetary exploration.

NASA was driven by the political goal of bridging the Soviet-U.S. "missile gap." The first two decades of NASA's existence clearly reflected this ethos, with the focus firmly on the manned Mercury, Gemini, and Apollo programs. JPL, under its visionary director William Pickering, though interested in the exploration of the planets, was responsible for the first series of U.S. moon probes, the Ranger series. These probes, as it turned out, proved to be the most stressful part of the relationship between JPL and NASA, because the first six *Ranger*s failed. That aside, the division of NASA/JPL labor generally worked well, particularly in later years when the planets came to the forefront of NASA's priority list.

One other player in this history, often forgotten yet arguably as important as JPL and NASA—particularly concerning the science and scientific rationale for planetary exploration—was the Space Sciences Board of the National Academy of Sciences (NAS). As early as 1957, the NAS recognized the importance of space, particularly with its involvement in the International Geophysical Year. Based on the successful International Polar Years of previous decades, the International Geophysical Year was a global commitment by scientists to study the physics of the Earth, including from space. America's launch of its first satellite, *Explorer 1*, in January 1958 marked the country's participation in the International Geophysical Year project. *Explorer 1*'s discovery of the Van Allen radiation belts helped convince the NAS to get involved in space research.

In 1958 the NAS created its Space Sciences Board (SSB) to define America's space science objectives. The committee was chaired by the distinguished geophysicist Lloyd Berkner and contained other luminaries such as Nobel Laureates Harold Urey and Joshua Lederberg and the incomparable planetary scientist Carl Sagan. An early and major preoccupation of the SSB was the question of extraterrestrial life, and it quickly established the Panel on Extraterrestrial Life to examine the problem. The panel initially consisted of two parts, EASTEX and WESTEX, named for their locations on the eastern and western seaboards, respectively. WESTEX was chaired by yet another Nobel Laureate, Melvin Calvin, whose work on photosynthesis had led to the identification of the Calvin cycle, a staple of high-school biology textbooks; EASTEX was led by Lederberg.

Lederberg was interested in discovering extraterrestrial life, but he was also focused on the problem of contamination. Lederberg was concerned both about back-contamination—alien microbes brought back to Earth by crewed missions—and forward-contamination—the release of Earth microbes onto other worlds that could threaten extraterrestrial life or be a potential contaminant when searching for evidence of extraterrestrials. The establishment of the SSB and the Panel on Extraterrestrial Life had profound implications for the future of space exploration, because it meant that the NAS would, in effect, set the scientific agenda for the investigation of the planets.

In 1960, EASTEX and WESTEX were merged with a group from the armed forces (also chaired by Melvin Calvin), forming the National Research Council Committee on Bioastronautics. This committee became known as Panel 2 on Extraterrestrial Life, and it was a very different kettle of fish than the SSB's subcommittees. Panel 2's membership included military men who were much less interested in extraterrestrial life than in questions of spaceflight, specifically whether there would be problems putting men in space. Yet Panel 2 created a subcommittee, chaired by Lederberg, called Committee 14 on Exobiology, which had the sole purpose of investigating extraterrestrials. This step was quite significant because it marked the creation of a new field of science: exobiology or, as it is also known, astrobiology, the study of extraterrestrial life.

In 1959 the very first grant to investigate extraterrestrial life was awarded to Committee 14, headed by Wolf Vishniac, a founding member of EASTEX. The $5,000 award was to develop "a prototype instrument for the remote detection of microorganisms on other planets," the first step in the long road to the Viking missions. Committee 14's new project became known as the "Wolf Trap."

Vishniac's team struggled for an adequate practical definition of life (as opposed to the theoretical definitions of life described in the last chapter, which are much harder to test for) on which to base their tests. They concluded that there were essentially four categories into which life on other planets might fit. First, life could be carbon based, like Earth's, for which the standard protocols for detecting carbon-based organisms (which date back to Louis Pasteur) could be employed. Second, extraterrestrial life might be based on Earth-like chemistry but with its own peculiarities and idiosyncrasies resulting from its evolution in a different environment. This contingency would allow standard chemical techniques to be used for detection. Third, extraterrestrial life might never have left the microbial stage and would require new, particularly sensitive techniques to find it. The fourth, final, and most worrying category of all was that life on another planet might be based on an entirely different chemistry than Earth's carbon chemistry, making chemical characterization virtually impossible.

As an alternative to carbon-based life forms, Committee 14 considered silicon, which, with its proximity to carbon on the periodic table, shares many properties with it. The reactivity of an atom is a function of the number of electrons orbiting in its outer shell. These electrons determine the number of bonds that can be formed with other atoms to make compounds. This ability to diversify is what makes carbon such a prevalent component in life-based organisms. Silicon, like carbon, can form a wide range of more complex chemical compounds. A problem with silicon-based life is silicon's affinity for oxygen. For a carbon-based life form on Earth, the oxidization of carbon during respiration forms the gas carbon dioxide—a waste product easily disposed of through exhalation. The oxidation of silicon produces a solid, silicon dioxide, which cannot be easily eliminated.

The idea of life based on silicon is not new. In 1891, Julius Shiner, an astrophysicist at the University of Potsdam, speculated that life based on the silicon atom might be common in the universe. The idea was elaborated by James Emerson Reynolds who, only a couple of years later, pointed out that the stability of silicon might allow life to exist at high temperatures. Three decades later, the brilliant British biologist and essayist J. B. S. Haldane suggested that iron-oxidizing, silicon-based life might be found deep within planets' interiors.

All these hypotheses were based on the assumption that life exists on other planets, an assertion Committee 14 was wary of. Furthermore, if silicon-based life could exist, what other atoms might life potentially be based on? The possible list was enormous, which led Committee 14 to make clearer delineations for what we do and do not consider to be a life form. The first, and most important, of these principles was that all life is based on the chemistry of carbon. This idea permeates all of NASA's astrobiology initiatives to this day and may or may not be correct. It is, however, a reasonable assumption because of carbon's extreme reactivity. After all, the entire field of organic chemistry exists just to study the multitude of compounds carbon can form.

## THE LABYRINTHS OF MARS

Mars has been a target of interest since the mid-nineteenth century when Giovanni Schiaparelli turned his telescope on its enigmatic red disk. Schiaparelli's fame springs from the maps of Mars's surface that he made with his telescopic observations. With this simple act, Schiaparelli forever changed the face of, and the nature of our relationship with, the red planet; he claimed the planet for humanity no less than Lewis and Clark claimed the American West for the United States.

In 1877 viewing conditions for Mars were their most favorable for thirty years, and the planet consequently attracted the attention of observers all over the world. It was the planet's particularly favorable opposition that year that inspired and enabled Giovanni Schiaparelli to "geographize" it, and in the process, he named many of the planet's topographic features, making the red planet seem close, accessible, and, to an awed Victorian public, very possibly inhabited. In labeling Mars's features, Schiaparelli famously broke with the prevailing conventions of astronomical naming, which, at the time, was to use classical terms drawn from Greek. Instead he used romantic-sounding place names taken from a wider spectrum of Latin and Mediterranean mythological literature. He thus named the northern Martian plain (where, more than one hundred years later, *Beagle 2* would be lost) Isidis Planitia and the huge impact crater dominating the southern highlands Hellas Planitia. A prominent, lighter-colored spot in the southern hemisphere he named Nix Olympia—meaning the snows of Olympus—which today we know of as Olympus Mons, the largest volcano in the solar system. He named other light patches Elysium, Thyle, Tharsis, and Cydonia, while a great triangular patch that had first been observed by Christian Huygens over two hundred years earlier—and that Huygens thought was a marsh—Schiaparelli named Syrtis Major.

Although Schiaparelli named over three hundred Martian features, he is perhaps most remembered for naming the features that he thought connected many of these light and dark patches. Schiaparelli named these strange, linear features *canali*, the Italian word for channels—features usually attributed to natural processes. The correct meaning was lost in translation when Schiaparelli's "channels" were translated literally as "canals," implying a feature made by humans or, in this case, aliens.

The channel/canal confusion set the stage for Percival Lowell, Schiaparelli's most famous disciple. When Lowell entered the Martian debate in the last years of the nineteenth century, a belief in Martian life was common. But it was Lowell in particular who misinterpreted Schiaparelli's *canali* and construed the linear features on the planet's surface as true canals; this became evidence not just of life but of advanced life. With disarmingly simple reasoning, Lowell suggested that the presence of complex structures implied the presence of engineers and the fact that they were canals showed the need for irrigation, which implied a water shortage. That, coupled with the fact that there was no obvious sign of extant engineers, led Lowell and his contemporaries to conclude that the society that had built the canals was gone and that their planet was dying. This jelled with the idea that Mars was older—and thus further along in evolution—than the planets closer to the sun.

## THE BICENTENNIAL MEN, PART II

By 1960, the deliberations of Committee 14, in conjunction with JPL and NASA, were focused on Mars as the most obvious place to search for extraterrestrial life. The legacy of Schiaparelli and Lowell was hard to shake and, with the euphoria surrounding the imminent manned Mercury missions, NASA began planning a series of missions with sophisticated probes to Mars. The first missions would be flybys, followed by an orbiter mission and culminating with a spacecraft that would land on the surface. It was an amazingly ambitious and carefully crafted program.

There were two stumbling blocks in these plans, the first being that the boosters available at the time were inadequate to launch a craft that would reach the red planet. The second problem was even more profound: How would NASA search for life on another planet?

The problem with the launch vehicle was simply a matter of technology. In 1961, first a chimpanzee named Ham, then Alan Shepherd and Gus Grissom, successfully rode the Mercury-Redstone vehicle into suborbital flight. But the Redstone was not powerful enough to launch a capsule into orbit. Partly compelled by Russian Yuri Gagarin's successful Earth orbit, NASA developed the more powerful Atlas rocket to launch astronaut John Glenn into Earth orbit. Bob Gilruth, Glennan's successor as NASA administrator, was fully aware of the extreme risk he was asking Glenn to take with a rocket that was not fully certified as safe. Despite the dangers, Glenn's mission aboard *Friendship 7* was a stunning success and proved the Atlas launch vehicle reliable.

The planned flyby and orbiter missions to Mars required a first-stage Atlas rocket coupled to a powerful second stage. The design for the second-stage rocket, called Centaur, used a highly explosive combination of liquid hydrogen and liquid oxygen to deliver far more thrust than any previous booster. Confidence in the Centaur rocket was so high that, by 1960, NASA and JPL were already discussing the design of a Mars landing mission, which they named Voyager. But the development of Centaur took longer than the designers expected, which had an immediate and lasting impact on the scheduling of all planned missions. Eventually, in 1964, the Atlas-Centaur rocket was certified, and *Mariner 4* was launched to the red planet.

*Mariner 4* arrived at Mars on July 15, 1965. At its closest approach, it was a little more than six thousand miles away, and it returned the first "close-up" pictures of Mars. *Mariner 4* took twenty-one complete photographs and twenty-one lines of a twenty-second picture, transmitting a grand total of 5.2 million bits at a data rate of 8.33 bits per second, the fastest its onboard data storage device, a tape recorder, could handle. *Mariner 4*'s data trickled in to JPL so slowly that the waiting scientists had time to print out each pixel's value and mark it on a large chalkboard to create an emerging mosaic of Mars, adding one new square every two and a half minutes. To provide some perspective, today's Deep Space Tracking Network consists of three radio telescopes in California, Australia, and Spain that handle data streams of up to 110,000 bits per second, and a standard household broadband modem downloads at a rate of several million bits per second. We have come a long way since *Mariner 4*.

Given the poignancy of the moment, it was a pity that the picture emerging on the JPL chalkboard was not one anyone expected to see. The pictures showed a heavily cratered terrain more reminiscent of the moon than of the Earth. The craters implied that Mars was tectonically lifeless, which seemed to spell the end to any ideas that Mars could be or once was a habitable planet. The theories of Schiaparelli and Lowell were exposed as pure myths, leaving scientists questioning the merits of sending life-detecting ships to Mars at all.

Gloom settled over the Mars exploration community with the depressingly bleak vision of Mars returned by *Mariner 4*. Not only did the pictures show a planet that looked far too similar to the cratered and lifeless moon, but the ingenious occultation experiment, which measured changes to radio waves sent from *Mariner 4* to Earth through Mars's atmosphere, indicated that the atmosphere was much thinner than had been estimated. Optimists had claimed that the air pressure on Mars could be up to one hundred millibars—a tenth of the average sea-level atmospheric pressure on Earth. The occultation experiment's results were unequivocal: the ground-level atmospheric pressure on Mars was less than ten millibars. Scientists had also debated for centuries what the Martian polar ice caps were composed of. The debate was definitively decided when the occultation experiment revealed that Mars's atmosphere is composed mainly of carbon dioxide.

Despite these disappointments, two further probes, *Mariners* 6 and 7, were ready for launch in 1969 and arrived in Mars orbit only a few days after Neil Armstrong and Buzz Aldrin landed on the moon. The journey had not been without incident. Just a few hours before it was due to fly by Mars, *Mariner 7* fell silent. The consensus at JPL was that the probe had been hit by a meteor hard enough to disrupt its orientation lock on the star Canopus. Luckily, however, it was still able to receive commands transmitted from JPL, and eventually the spacecraft was persuaded to re-lock onto its orientation star. Although the spacecraft had been damaged by the meteor strike, its trajectory changed so little that its encounter with Mars was only ten seconds later than had been scheduled. Ironically, of the two spacecraft, it was the damaged *Mariner 7* that provided the better pictures.

Both probes were still a long way from Mars when they started taking pictures, which showed the southern polar cap in great detail as well as the huge structure Schiaparelli had seen and immortalized in his maps—Nix Olympia or Olympus Mons, a giant volcano fifteen miles high. There was no sign of Schiaparelli's *canali,* Lowell's canals; it was the final nail in the coffin for an idea that had persisted for almost a century.

Both probes confirmed Mars's low atmospheric pressure and found that the mean daytime temperature at the equator, about 60 degrees F, dropped to a bone-numbing low of –100 degrees F at night. *Mariner 7*'s survey of the southern polar region showed that the temperature was a staggering –190 degrees F, close to the temperature at which carbon dioxide turns to ice. Many believed that this finding confirmed that the Martian ice caps are composed of solid carbon dioxide. *Mariner*s *6* and *7* also discovered a highly distinctive double crater, which was quickly named the Giant's Footprint.

Together the two probes mapped about 20 percent of Mars's surface and showed three types of terrain: cratered areas, areas with few craters but many jumbled ridges (dubbed "chaotic terrain"), and infrequent smooth areas of which the Hellas Basin is the best example. This mixture of terrains was what inspired Steve Gould to consider Mars in an intermediate state between tectonically active and tectonically dead.

To the JPL engineers, *Mariner*s *6* and *7* seemed to broadly confirm the findings of *Mariner 4* that Mars's surface is overall similar to that of the moon. The dark areas that seemed from Earth-based telescopes to wax and wane with the Martian seasons were not found in the high-definition pictures returned by the probes. Furthermore, the seeming confirmation that the polar caps were composed exclusively of carbon dioxide ice, coupled with the fact that the atmosphere was almost exclusively gaseous carbon dioxide, reinforced the impossibility of life on the red planet.

Regardless of JPL's disappointment at this evidence, *Mariners 8* and *9* were already in production. These Mariner missions were tasked with photographing potential landing sites for the planned Viking landers scheduled to land during the 1976 bicentenary year. In all, four probes were slated for launch during that 1971 window: the two American Mariners and two Russian probes, *Mars 2* and *3*. Of the four probes, three had extremely short life spans. *Mariner 8* crashed into the Atlantic some 350 miles northwest of Puerto Rico after its second-stage Centaur rocket failed to ignite. *Mars 2* and *3* made it to the red planet but quickly failed.

The Russian probes were ambitious, consisting of orbiters and landers. The first to arrive, *Mars 2,* entered orbit on November 17, 1971, and ejected a capsule that crashed into the Martian surface about three hundred miles southwest of the Hellas Basin. No scientific information was returned, and its main claim to fame is that it was the first manmade object to make contact with the surface of the red planet. *Mars 3* also ejected a capsule, which safely soft-landed not far from the northern limit of the southern polar ice cap. Within ninety seconds of arrival, it began transmitting a picture of the Martian surface. After only twenty seconds, however, transmission broke off and was never re-established. To this day, no one knows what happened, although there was some speculation at the time that the parachute that carried the lander to the surface settled on top of the craft! If so, it was a disaster in the same vein as the one that befell the Russians during their *Venera 14* mission, when the craft's spring-loaded sampling arm landed on its lens cap, prohibiting it from taking any measurements.

In 1971, however, one of the worst dust storms in living memory was raging across the Martian surface. Scientists had known for a century that such dust storms were particularly common when Mars is in opposition—at its closest point—to Earth. It could have been this dust storm rather than the parachute that prevented the *Mars 3* lander from following through on its mission. At any rate, at least both the *Mars 2* and *3* orbiters worked well, and they sent back useful information that confirmed estimates of Martian atmospheric pressure, as well as surface topography.

The real success of the 1971 suite of missions was *Mariner 9*, the sister ship to the ill-fated *Mariner 8*. *Mariner 9* became the first probe ever to orbit another planet (although it was a while before she sent back any useful information because of the raging dust storm). *Mariner 9* beat the Soviet *Mars 2* and *Mars 3* by only a narrow margin, for all three arrived within a month of each other. Toward the end of the year, the dust began to clear, and *Mariner 9* got its first view of the Martian surface. The four dark patches that it had faintly seen through the year-long gloaming turned out to be the summits of Olympus Mons and the three huge volcanoes of the Tharsis range: Ascraeus Mons, Pavonis Mons, and Arsia Mons.

*Mariner 9* put a smile back on the faces of the JPL scientists and engineers by showing that, contrary to what they had come to expect, the surface of the red planet is extremely varied. *Mariner 9* showed a previously undreamed of vista of high mountains, deep valleys, and wide basins. Towering above it all was the gigantic, unbelievable bulk of the fifteen-mile-high Olympus Mons. During the *Mariner 9* mission, scientists realized that a sea-level datum-line on this world without seas would be required. The baseline, in the absence of any other data, had to be based on atmospheric pressure and was accordingly placed at 6.2 millibars.

But *Mariner 9*'s lasting claim to fame was in illuminating the difference between Mars's southern and northern hemispheres, quite in contrast to what previous expeditions had shown. They are so distinct that they could belong to two different planets. The southern hemisphere is heavily cratered, with an average altitude of about a mile above the datum-line. The terrain of the northern hemisphere is significantly

Color composite of the sun from the SOHO project.

*All photos courtesy NASA/JPL-Caltech*

*Mariner 10*'s first image
of Mercury, March 24, 1974.

Venus centered
at ninety degrees
east longitude.
Color coded to
show elevation.

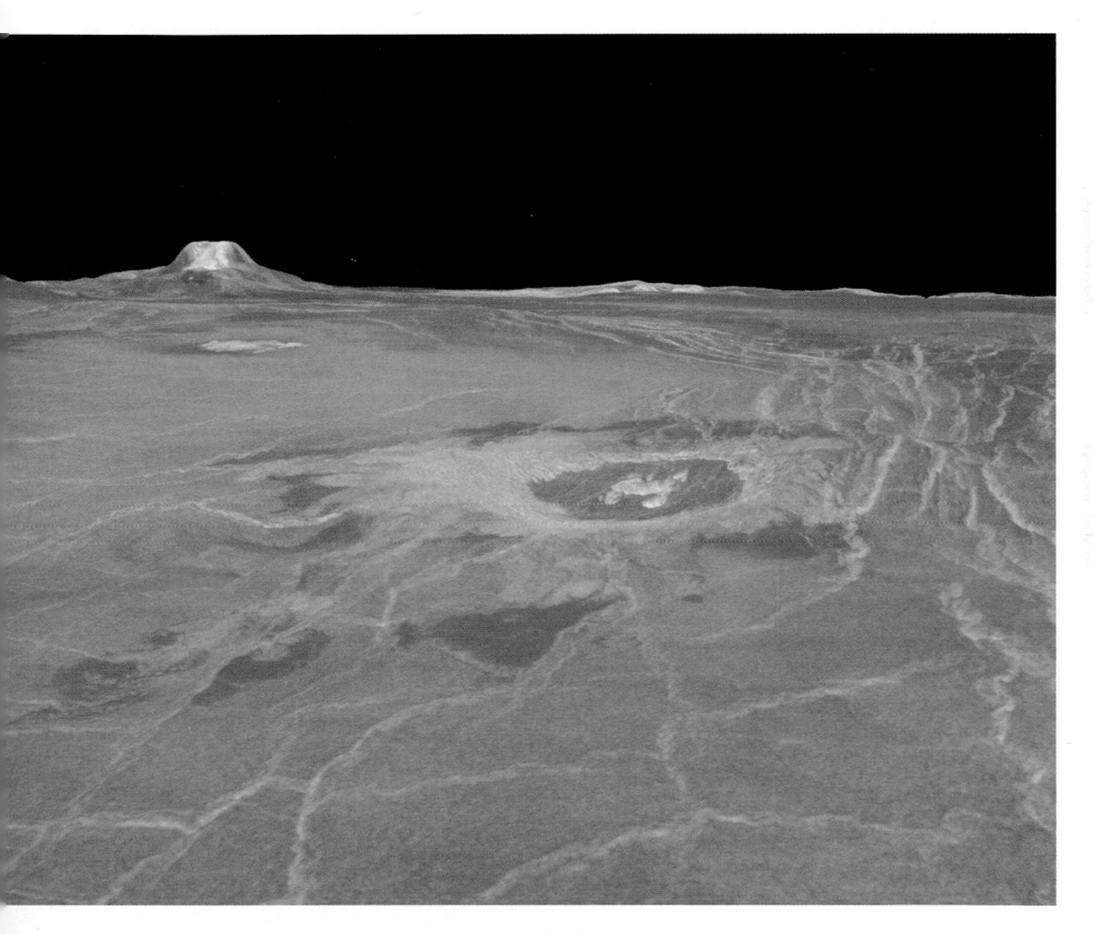

3-D perspective view of the Eistla Regio area of Venus
with the volcano Gula Mons in the background.

The McMurdo panorama taken by the Mars Exploration Rover, *Spirit*. The groove dug by *Spirit*'s broken wheel is clearly visible.

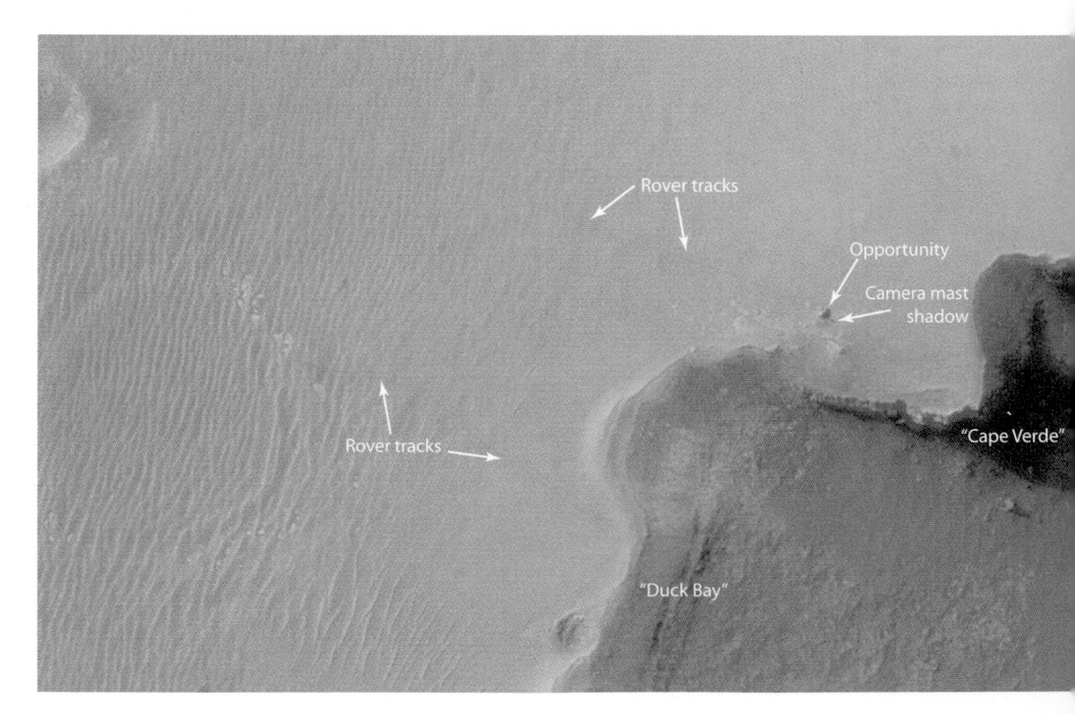


*Opportunity* at the edge of Victoria crater seen from the Mars Reconnaissance Orbiter.

*Galileo* image of
Callisto showing its
eavily cratered surface.

Jupiter and Ganymede
imaged by the passing
*Cassini* spacecraft
en route to Saturn.

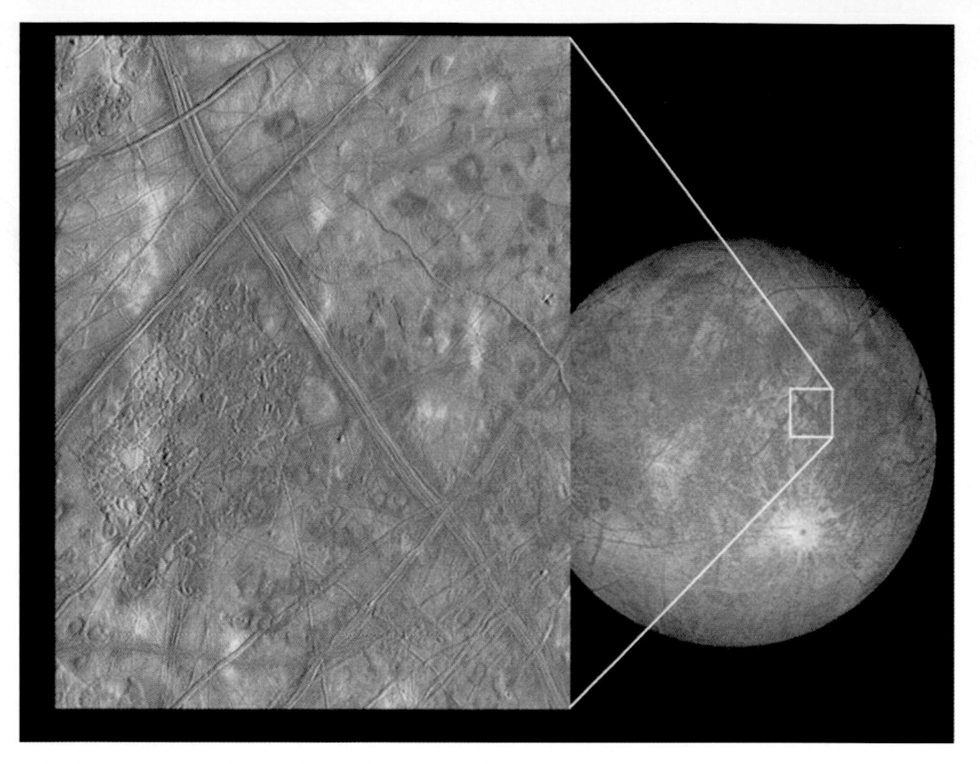

Blocks of ice on the surface of Europa which have broken apart and reformed are evidence for the presence of a subterranean ocean. Taken by the *Galileo* orbiter.

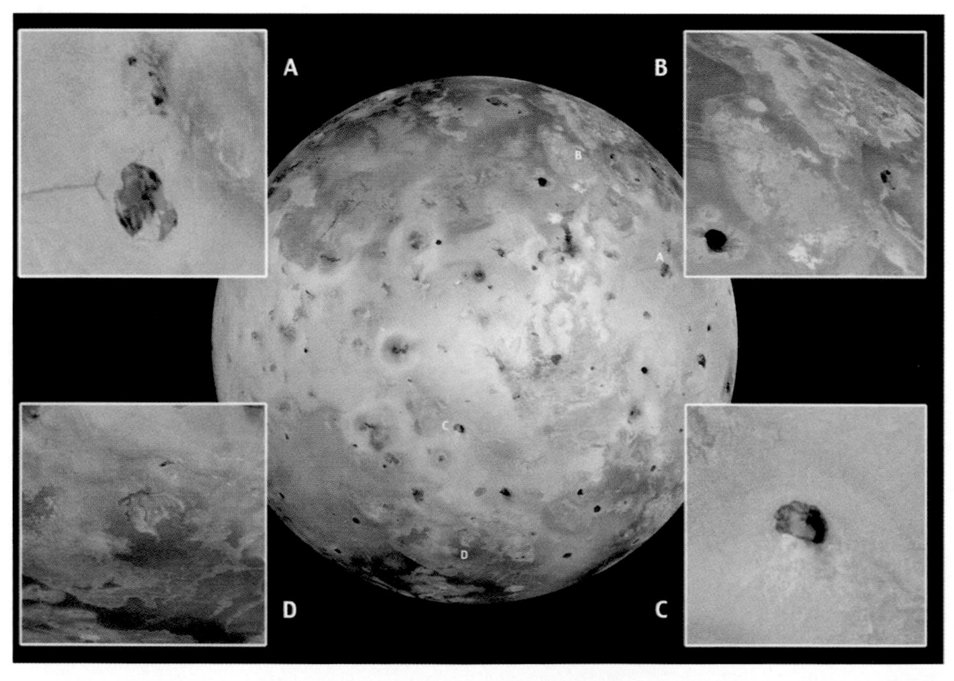


Images of Jupiter's volcanic moon Io obtained by the *Galileo* spacecraft in 1999. A, details of lava and sulfurous deposits; B, details of whitish high-latitude deposits; C, fissures on the surface; D, sulfur flows.

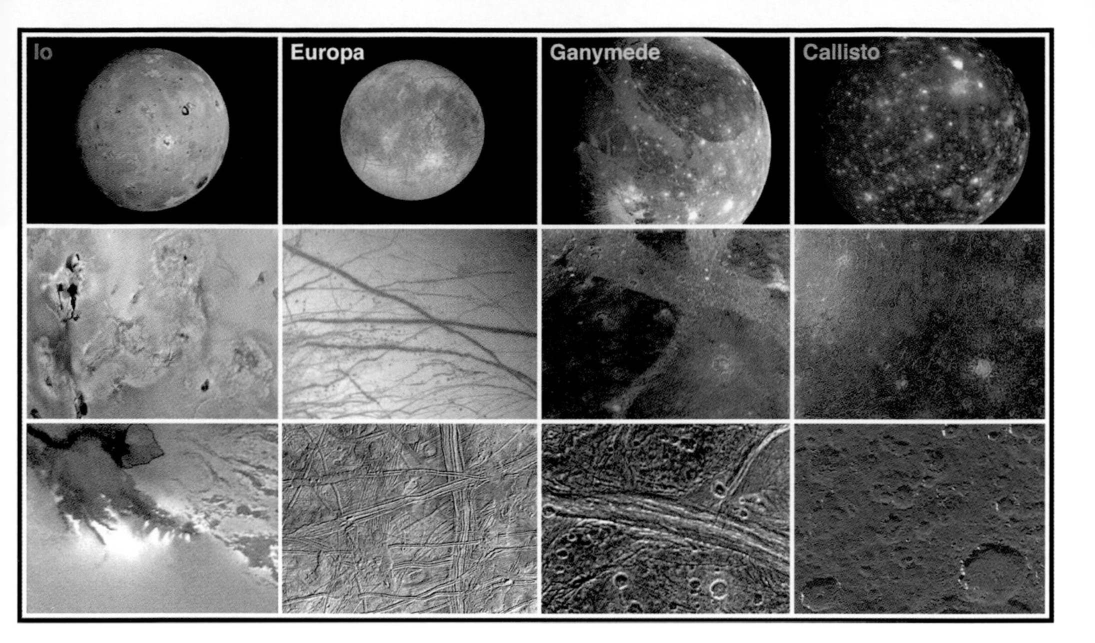


*(Above)* Postcard from *Galileo*. Nothing illustrates the diversity of the Jovian moons more than their different surface textures, which in turn reflect their varied geologic histories. The top row shows the moons at the same scale with a surface resolution of about 20 kilometers. The middle row shows surface features with a resolution approximately ten times higher. Io's volcanic craters are visible, the tidally induced cracks on the surface of Europa, Ganymede's brightly grooved surface, and Callisto's giant impact basins. The bottom row shows still higher-resolution pictures of the moon's surfaces where features as small as 20 meters across have been resolved.

*(Left)* Family portrait of Jupiter and her major moons. From top to bottom the moons are Io, Europa, Ganymede, and Callisto. Jupiter's great red spot is a storm that is at least three hundred years old.

Titan imaged from the approaching *Cassini*.

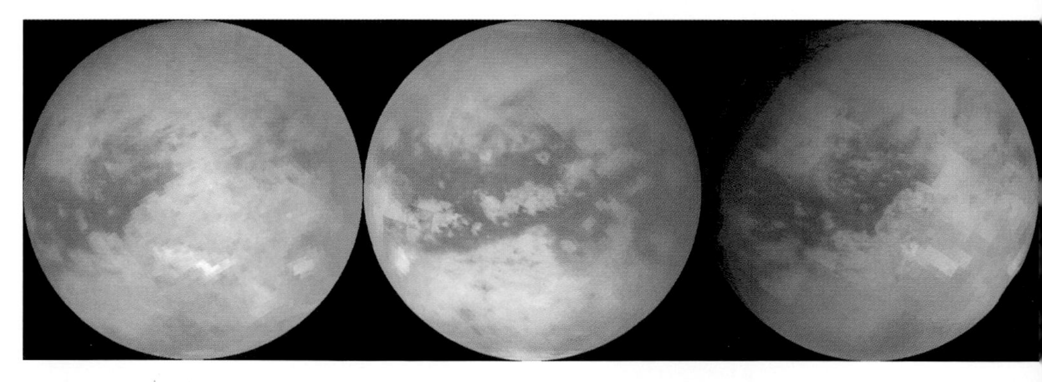

Mapping Titan's surface. The middle image is the opposite hemisphere to the images on the right and left. Note the bright clouds at the south pole in the middle image which are not visible in the other images. This suggests that Titan's atmosphere is in a state of constant flux like our own. The right and left hemispheres show an equatorial tick-shaped mark (known as the Chevron) that is thought to be a surface deposit of water or carbon dioxide thrown out from one of Titan's many volcanoes.

Artist's impression of *Huygens* landing on Titan.

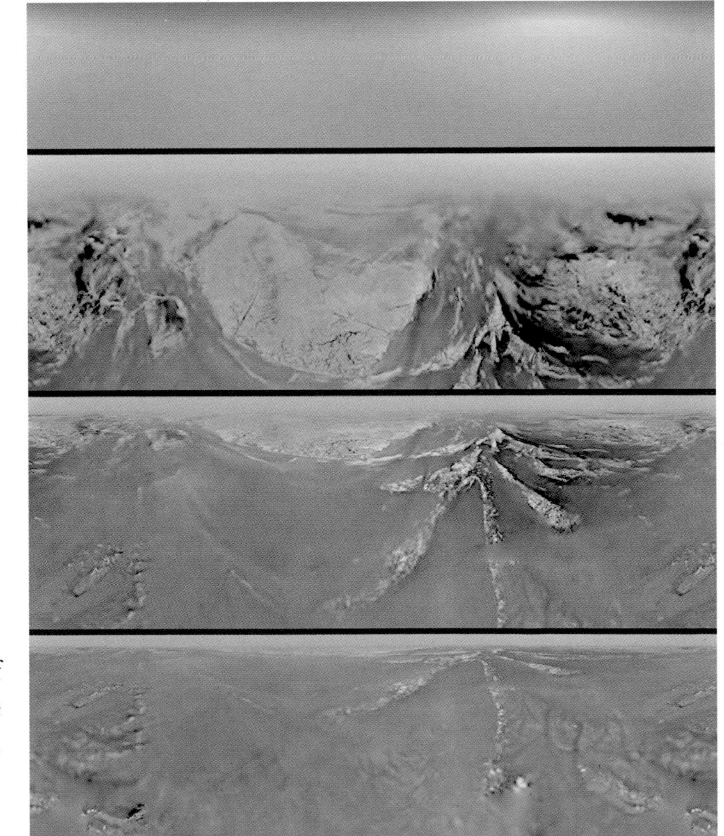

Emerging vista of Titan during the descent of *Huygens*.

Artist's impression of the site of *Huygen*'s landing.

Saturn's moon, Mimas.

The Rings of Saturn taken by *Cassini*.

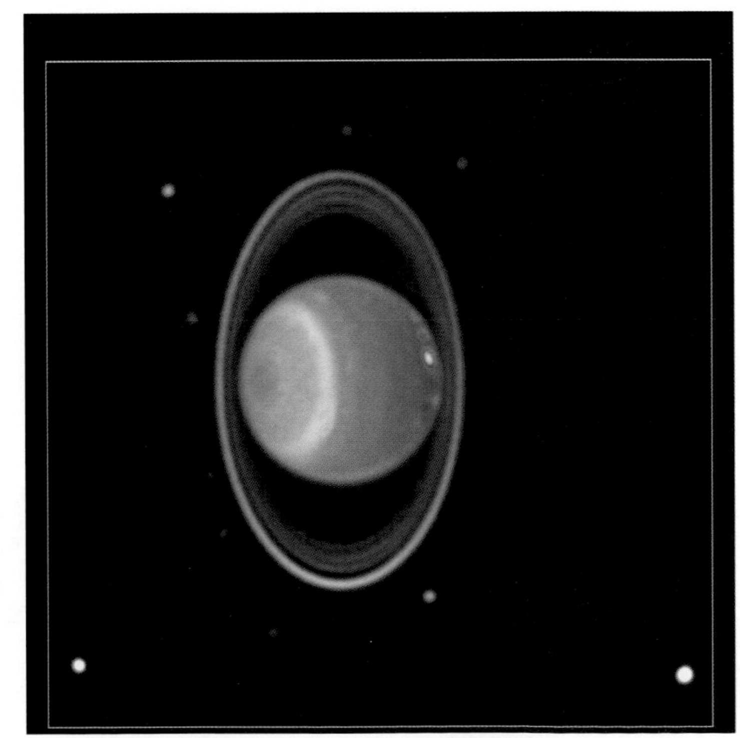

Uranus, its rings and some of its satellites imaged by the Hubble Space Telescope.

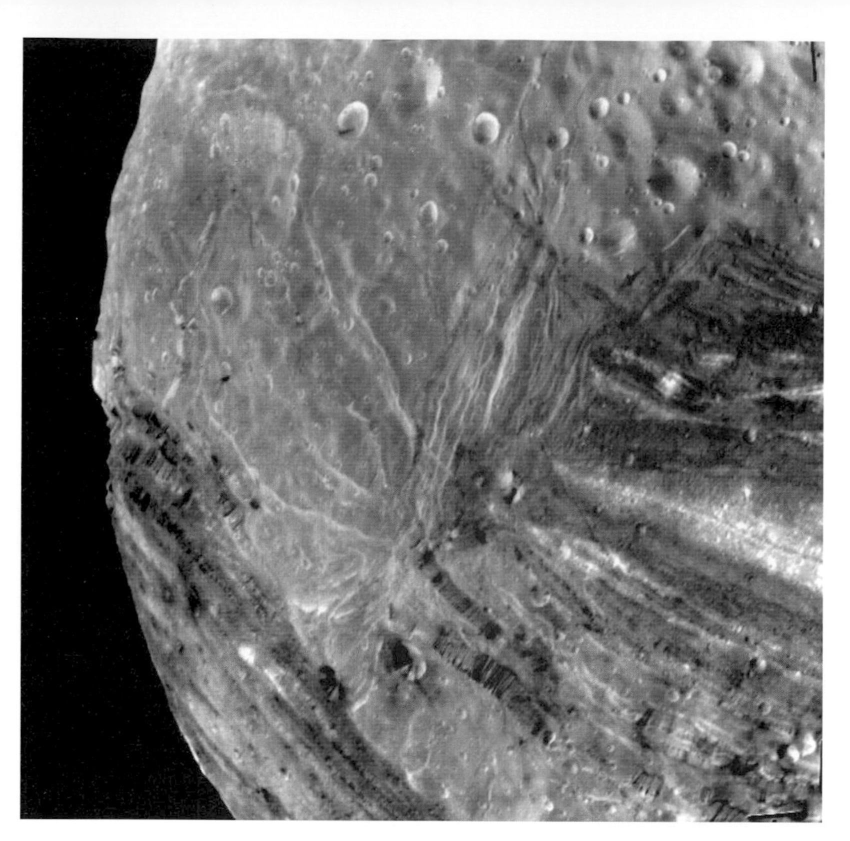

Surface of Uranus' tormented moon: Miranda, taken by *Voyager 2*.

Oberon, outermost moon of Uranus. Taken by *Voyager 2*.

Neptune seen from behind its most famous moon, Triton.
Computer-generated montage of *Voyager 2* images.

Artist's impression of Charon seen from the surface of Pluto.

Kuiper Belt objects, Sedna and Quaoar, compared with Pluto, the moon, and Earth.

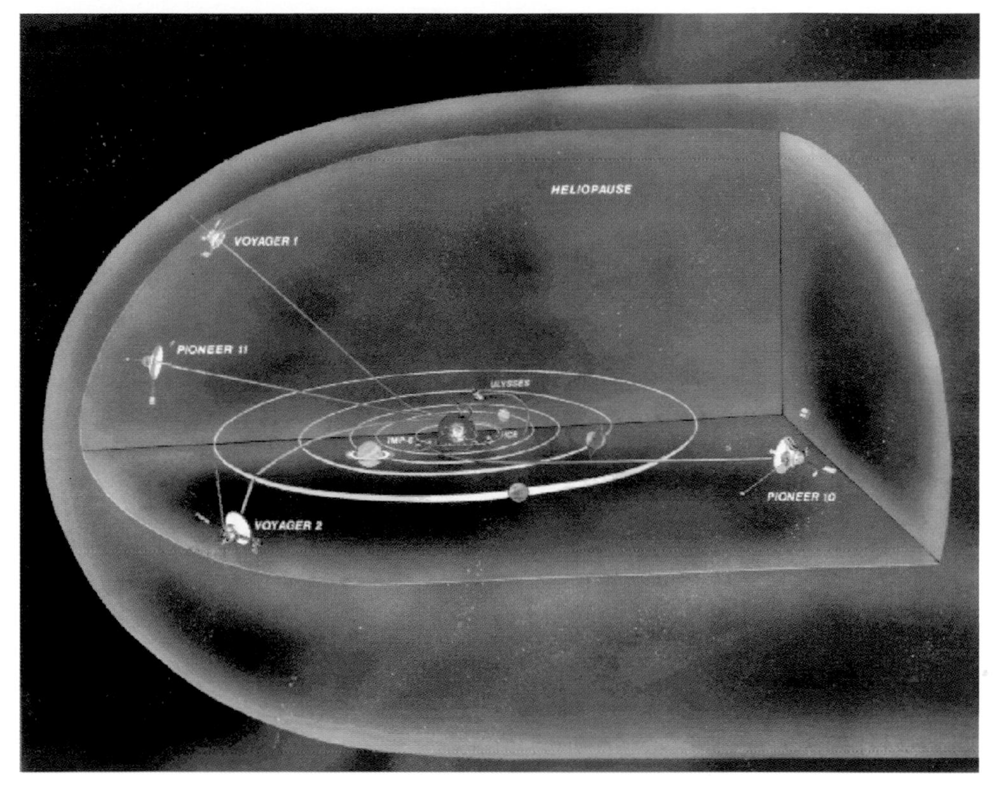


Approximate positions of the *Voyager* and *Pioneer* spacecraft relative to the solar system and its edge, the Heliopause.

Artist's conception of a view from a hypothetical moon orbiting the planet HD 188553 Ab, 149 lights years away in the triple star system HD188753 in the constellation Cygnus.

Artist's conception of a *Voyager* spacecraft entering interstellar space.

lower—mainly below the 6.2-mb datum-line—and much more lightly cratered. The implication was immediate and obvious to the scientists at JPL: the southern hemisphere is old, and the northern hemisphere, young. And yet, the division between the northern and the southern hemisphere terrains is weirdly skewed, with the dividing line between them at a thirty-five-degree angle to the planetary equator.

## THE BICENTENNIAL MEN, PART III

*Viking 1* launched on August 20, 1975, from Pad 41 at Cape Canaveral at the end of a week that had not been, by any standards, easy. Problems had required that the spare spacecraft be substituted for the primary one. (This era was the age of big science in planetary exploration—the Viking budget, at an eye-watering $1.1 billion, was large enough to include the construction of a spare spacecraft.) *Viking 2*'s launch on September 9, 1975, was, if possible, even more tense than *Viking 1*'s. Shortly before lift-off, with bad weather gathering over the Cape, NASA's meteorologists sent launch control a message that the spacecraft had to depart within the next ten minutes; any later, and they would have to scrub. The timing was almost impossibly tight; the slightest hitch, and *Viking 2* would miss her launch window completely. But *Viking 2* successfully launched at 2:39 PM, just three minutes before the scrub order would have been given.

Both Viking craft were huge, weighing in excess of three thousand kilos, and they required the power of the massive Titan III-E rockets and the newly de-bugged Centaur for the final boost into deep space. Each spacecraft consisted of two components, an orbiter and a lander. The orbiters contained imaging and scientific experiments, while the landers carried the instruments, particularly the life-detecting experiments that would be used on the surface. These life or biology experiments were marvels of miniaturization, with each weighing less than sixteen kilos while containing three subexperiments: the pyrolytic release, the labeled release, and the gas exchange experiments. As an adjunct to the biology package, the Viking missions also carried a gas chromatograph/mass spectrometer (GCMS) to measure the composition and abundance of any organic compounds the spacecraft might find in Martian soil. Organic means "carbon-containing," and the inclusion of the GCMS reflected the Viking team's underlying conviction that any life on Mars would be based on the carbon atom.

The labeled release experiment was designed to moisten a small sample of Martian soil with a controlled amount of nutrient culture medium. The organic compounds were labeled with radioactive carbon–14 so that, after a suitable incubation period (about ten days), any microorganisms would have consumed the labeled nutrient compound and would be giving off waste gases containing the carbon–14, which could then be detected. On Earth, terrestrial organisms release carbon dioxide, carbon monoxide, and methane, all of which show up labeled with carbon–14 if it is added.

The gas exchange experiment added a complex brew of chemicals (which the Viking experimenters called, with a fine sense of irony, "chicken soup") to a small sample of Martian soil. After the chicken-soup dunking, the soil sample would be incubated for at least twelve days in a simulated Martian atmosphere of carbon dioxide, with added inert gases such as helium and krypton. Any gases produced by Martian organisms consuming the nutrient would be detected by a gas chromatograph.

Hindsight is wonderful, and from what we now know about Mars, one problem with these experiments is immediately apparent. They both relied on the use of liquid water, a commodity that is not, as we have discovered, common on the surface of Mars. Only the third Viking biology experiment—the pyrolytic release experiment—did not use liquid water and so actually came close to simulating the nature of the surface of Mars. In this experiment, a tiny sample of Martian soil was incubated in a simulated Martian atmosphere of carbon monoxide and carbon dioxide labeled with carbon–14. A xenon arc lamp provided artificial sunlight. After five days, the atmosphere was flushed away, and the sample was roasted to *destroy* any organic material. The resulting gases were then tested to see if any organism had taken up the carbon–14-labeled atmosphere.

Despite these refined experiments, it was the GCMS that really answered the question of Martian life, and it found no traces of any organic compound. The result stunned the Viking scientists, because organic molecules were already known to be in the deepest pockets of interplanetary space, having been discovered by scientists studying meteorites. Calibration checks showed that the machine was working perfectly; there was no question about its results. The lack of any organic molecules made a mockery of the other three biology experiments, each of which had been designed to hunt for the spoor of complex organic life—respired gases, waste products, and so on. Without organic molecules, the three sophisticated biology experiments were just dead weight.

Yet the biology experiment results were puzzling. Each contained a "control" to prevent false positives—not only did each experiment need to find life, it also had to *not* find it in identical soil samples that had been sterilized by heating. Both the gas exchange and the pyrolytic release experiments produced positive results, even though the control samples for both experiments and the lack of organic compounds demonstrated by the GCMS indicated that no biological processes could possibly have been at work. Subsequent laboratory experiments on Earth showed that Mars had probably thrown a curved ball at Viking; highly reactive oxidizing compounds that were lying dormant in that world's red soil would, when exposed to water—the universal catalyst that the Viking scientists had loaded their spacecraft with—produce hydrogen peroxide. Copious amounts of oxidized iron, the very material that gives Mars its red color, could, for example, easily have produced the results seen by the pyrolysis experiment.

Only the labeled release experiment held out any hope, and even that hope was short lived. Upon initial injection of the nutrient, the gas released from the sample increased. Further injections caused the amount of gas released to decrease and then increase once again. No response was seen from the control sample. The general feeling was that the labeled release results could also be explained by nonbiological processes.

Eventually, the Viking team, including its most charismatic member, Carl Sagan, would write in the *Journal of Geophysical Research* in 1977 that, as far as they could see, there was no life on Mars. It appeared to be the end of the road for the hunt for life on the red planet.

The Viking mission cost $1.1 billion and has to stand as one of the most noble failures in the history of space exploration, if only for the optimism—and perhaps naivety—with which it was dispatched. When the Viking mission was conceived and launched, the science of astrobiology barely existed, even in the loose and axiomatic form it takes today. There was no one to say what was the right or the wrong way to hunt for life on another planet, and the Viking designers gave it their best shot—at a time when space dollars were literally running in the gutters. To send a spacecraft to search for life in the 1970s using mainly water-based experiments was, to our eyes now, clearly a mistake. It was a bold—and perhaps rash—move; yet, it rode a wave of optimism that had accompanied the Apollo missions with all their glory, and what right do we have to gainsay it now?

Since Viking, NASA has spent millions developing astrobiology to a state where we can properly ask the kind of questions that we sent the Viking spacecraft to answer. We are getting closer, but the questions, and their answers, are still awfully complicated. If we were to honestly assess the heroic age of the Viking landers, we would have to say that the state of knowledge in the 1970s really could not successfully support such a mission. We tried to run before we could walk.

There were two major problems with the Viking missions: The landers were not mobile—they could not even look over the next hillock—and they were not equipped with the only experiment that will ever answer the question of whether there is life on Mars, for they did not carry a carbon isotope ratio mass spectrometer.

Despite the failure of a series of well-publicized Mars probes in the decades after Viking, NASA did at least get the message about mobility. Its *Pathfinder* lander in 1997 roved across the surface of Mars and showed that mobile Martian exploration could be done. That, in turn, paved the way for the Mars Exploration Rovers—the first robot geologists to visit another world. But the life experiment was a different story altogether that would only be re-visited three decades later by a British university professor and some of his friends, who, even fifty years after the end of the Second World War, still saw in their mind's eye the challenge of distant contrails as they ran for their scientific Spitfires. For Colin Pillinger and colleagues, the White Cliffs of Dover had been swapped for the sands of Mars.

## THE CARBON MACHINE

On December 25, 2003, I got up early and made my way downstairs past my sleeping household. I put a bottle of sparkling wine in the fridge to chill and turned on the television. The pictures being transmitted were from a studio in London, where the air of expectation was palpable. A hundred million miles away, a tiny, pocket-watch-shaped spacecraft was just about to transmit back the news that it had safely arrived at Isidis Planitia, a low plain-like area that looked like a chunk bitten out of the strangely tilted dividing line between Mars's northern lowlands and the southern highlands.

People I barely knew then, but would come to know well, were in that studio: Ian Wright, then a reader in planetary sciences at the Open University, and Colin Pillinger, the visionary scientist behind the *Beagle 2* project. In the background was my friend Stephen Wilkinson, a television producer with a passionate commitment to science communication who was there to cover one of the most momentous events in British space exploration.

We all waited, those in the London studio and me in the living room of our home in West Oxfordshire. I thought about the bottle of fizz that I would crack when the news came through that *Beagle 2* had successfully landed. But we all did a lot of waiting that morning. We knew, despite constant reminders from the UK Space Centre engineers, that retrieving a mobile telephone signal from another world was not a trivial task, that the longer we heard nothing the worse the news was likely to be and the greater the likelihood that *Beagle 2* had been lost.

And so that long morning wore on. Eventually I gave in and opened the bottle—it was Christmas after all—but by the time the family was stirring I was feeling what everyone in that television studio was feeling. *Beagle 2* was lost.

It was to have been the greatest success in British space history—plucky little Britain successfully achieving what the Americans with their vast industrial complex and their oceans of dollars could not do. We would find life on Mars. Never mind that the Americans had two hugely expensive spacecraft, the Mars Exploration Rovers, scheduled to land soon after *Beagle 2*. Neither of these rovers would look for life on Mars; both were robotic geologists that would answer whether there had once been water on the planet and would search for the existence of sedimentary rocks. These questions are undoubtedly good ones, but they are not the biggest question of all; they would not address whether we are alone in the solar system. Colin Pillinger named his spacecraft *Beagle 2* precisely because it was sent to search for life. The craft was to do for twenty-first-century biology what Darwin's voyage on the original *Beagle* had done for nineteenth-century biology: revolutionize it.

The singular instrument *Beagle 2* carried reflected this ethos, an instrument whose sole function was to test for the presence of life.

The instrument—smaller than a shoebox—was a stable isotope ratio mass spectrometer tuned exclusively to look for carbon. Built by Dennis Leigh in the backroom of an ordinary suburban, semi-detached house in the British midland city of Stoke-on-Trent, it was a masterpiece of miniature engineering.

Dennis was trained (as is almost all the mass spectrometer fraternity in the United Kingdom) within the hallowed halls of VG ISOGAS. VG (or GV as it is currently somewhat puzzlingly known) makes machines that weigh atoms. The company has gone through so many buy-ins, sell-outs, mergers, and acquisitions in its several decades that I sometimes think their engineers could have left and then rejoined without ever realizing it was the same company. The technique for weighing atoms goes way back to the 1950s and the Nobel Laureate Harold Urey, a man who, as previously mentioned, was passionately interested in the moon. Urey came out of the Manhattan Project with a full and deep awareness of the dangers of atomic physics, but he was also overridingly aware of the beneficial potential of this strange new science. He was anxious to see what it could do besides meet military goals, and his fertile imagination led him to examine how different stable isotopes can be exploited to yield information about their environment.

Isotopes are varieties of a chemical element (that is, atoms) that are chemically identical but physically distinct in that they have a different weight or, more accurately, a different mass. The difference in mass stems from the different numbers of neutrons in their nucleus. Neutrons are subatomic particles that have mass but no electrical charge. Urey knew that, if two or more isotopes are used in a chemical reaction, the end product of that reaction is not affected by the physical differences in the isotopes. Within the world of chemistry, there are two major groups of isotopes corresponding to the two principal types of elements: radioactive (unstable) isotopes and nonradioactive (stable) isotopes. Radioactive elements spontaneously transmute into other elements by the loss of alpha, beta, or gamma particles. Nonradioactive isotopes, being stable, do not transmute into other elements.

Having spent World War II fiddling with unstable isotopes, Urey's interest turned to the stable ones. He soon found that, because stable isotopes have a different mass, they are handled differently by purely physical processes; this difference—although fundamentally nonchemical in character—can be used to monitor physical processes.

Urey turned his attention first to the isotopes of oxygen. There are three stable isotopes of oxygen: the very abundant oxygen–16, the much less abundant oxygen–18, and the astonishingly rare oxygen–17, which we may safely forget about. Urey reasoned that, of the two former isotopes of oxygen, the lighter of the two (the one with the fewer neutrons, oxygen–16) would become more concentrated in a crystal precipitated out of a solution simply because its lesser mass allows it to move around more readily. The heavier oxygen–18 would less readily incorporate into the lattice of the crystal. The practical upshot of this finding is that, if you measure the ratio of the two isotopes in a crystal—say a crystal of calcium carbonate in a fossilized seashell—then you can determine the temperature at which it was deposited, that is, the temperature at the time the animal was alive. Urey's technique has worked staggeringly well in the fifty years since he invented it, and it has, in fact, become the basis of palaeoceanography, the science that studies the history of the Earth's oceans.

But oxygen is not the only stable element to have isotopes; another one is the element of life itself, carbon. Carbon has two stable isotopes, carbon–12 (with twelve neutrons in the nucleus) and the much rarer carbon–13 (thirteen neutrons in the nucleus). Unlike oxygen, the way in which these two isotopes behave, technically known as their fractionation, is not dependent on temperature. Rather their behavior is dependent on the one process that, on this planet above all, determines the distribution of carbon, photosynthesis. Just as the purely physical process of growing a crystal in ocean water prefers to use the light isotope of oxygen, photosynthesis prefers to use the light isotope of carbon, carbon–12, because it is easier to handle. When a plant grows, therefore, it preferentially concentrates the light isotopes of carbon, carbon–12, in its tissues.

As Committee 14 realized many years earlier, carbon is probably the basis of most, if not all, extraterrestrial life. The carbon test takes this reasoning a stage further; if carbon–12 is concentrated in a material, then the overwhelming likelihood is that this material is, or once was, living. There are good reasons to believe this assumption, and not just on purely theoretical grounds. As we saw in Chapter 4, some of the oldest rocks on Earth, 3.8 billion years old, exhibit an abundance of carbon–12, and this finding has been used to infer that life started while the Earth was very young, soon after the end of the late Heavy Bombardment Period.

Besides Dennis Leigh's isotope ratio mass spectrometer, *Beagle 2* carried a variety of equipment. The craft had a series of ovens designed to heat samples in stages to determine what component of the sample any measurable gas might come from. For example, carbon dioxide from carbonate is liberated at a much higher temperature than carbon dioxide from organic material. Another instrument was to determine the age of rocks using radioactive dating, another, different type of mass spectrometer was included for determining the mineral content of the rocks, and a suite of environmental sensors was to measure temperature and ultraviolet radiation at the surface. Also included in the craft were a barometer and detectors to register the speed and direction of dust particles and Martian wind.

*Beagle 2* also had eyes. It carried three cameras: a stereo pair to provide panoramic views of the landing site, a pop-up camera to provide a wide-angle picture of Mars, and a microscope camera to examine rock surfaces cleaned by the onboard grinder. To take its samples, *Beagle 2* was equipped with a mole. It was like something out of the movie *Thunderbirds*; after landing, the mole would crawl out of its stowage to inch its way across the Martian landscape. When it came to a suitable spot, it would stop and start drilling to get a fresh sample uncontaminated by Martian oxidation.

*Beagle 2* was as intricate as the pocket watch it resembled. But it was all to no avail. Two years after the tragedy of *Beagle 2*, the irony is that it looks as though it has been found, smack-bang in the center of its landing area. In that area is a crater, like a billion others on the surface of Mars. The orbiter *Mars Global Surveyor* has spotted the signs of a recent disturbance in this tiny crater. On its northern edge is a tiny dark discoloration that spills beyond the edge of the crater. For those skilled in interpreting pictures of other planets, this spot looks very much like the mark made by a small object when it strikes the ground traveling at high speeds. A few meters to the east is another small smudge that experts interpret as the first bounce mark as *Beagle 2* started spiraling down the crater walls like water down a plughole. Three marks nearby were supposedly made as *Beagle 2* spun to a halt. In the very center of the crater are a series of features arranged in a rough triangle that look exactly like the landing bag deployment of a nominal *Beagle 2* landing. A couple of meters to the east are four smudges, which could be the lander with its unfurled solar panels.

As I write, the best guess is that *Beagle 2* made it safely to the surface of Mars only to be captured by a crater on landing. In other words, the walls of the crater blocked its attempts to call home. Remember, its transmitter was only as powerful as a mobile phone. It would have kept sending until the last juice of its nickel-cadmium battery faded away, never receiving the "go" command to send the mole on its way and to fire up its intricate, magnificent, carbon machine.

This is just a scenario, of course. In many ways, it is reminiscent of another Martian tale—albeit of a different variety. That story is of ALH 84001, the Martian meteorite that in 1996 was purported to show evidence of life on Mars. That year David McKay and co-investigators at the Johnson Space Center published claims that a meteorite which—on the basis of its chemical composition—had come from Mars harbored evidence of life. In thc paper they published in *Science*, they highlighted four lines of circumstantial evidence to support this conclusion. They argued, first, that the carbonate minerals in ALH 84001 were formed at a temperature consistent with life, second, that organic chemicals (polycyclic aromatic hydrocarbons or PAHs) found in the meteorite were depleted in carbon–13 and therefore were traces of Martian biochemical reactions, third, that the magnetic grains in the rock came from bacteria (as they often do on Earth), and fourth—and most controversial of all—that strange structures found within the meteorite were the fossilized remains of bacteria. As was assumed by the Viking mission scientists, all four lines of evidence depended on life on Mars being organized along essentially the same chemical basis as on Earth. Although each line of evidence on its own did not prove anything, taken together they made the story of ALH 84001 highly suggestive.

Similarly, the story of poor *Beagle 2* contains various lines of evidence that stack up to a tale that describes its fate and could even be true. And there is, of course, the fact that these marks on the Martian surface were not there before the descent of *Beagle 2* but were there afterward, *and* they are in exactly the right place.

But there is a deeper truth here, which has to do with our need to make sense of the red planet. It is so far away and so very desirable that we try to impose our sense of order on it. We are desperate to understand it as a *place* rather than as just a dot in the sky, because a place is real and reachable and a dot is merely a dot. Like ALH 84001, the story of *Beagle 2*'s demise is another Martian fable woven out of our desire to touch that enigmatic red orb with our souls. And one thing is certain, as we reach closer and closer to the ultimate goal of humans standing on the red planet, other Martian myths will flower. The *Martian Chronicles* have only started to be written.

In November 2006, NASA's *Mars Reconnaissance Orbiter* took up position in its parking orbit above the red planet. The pictures so far are stunning, and the orbiter's chief scientist has promised to turn his craft's camera on Isidis Planitia to try to find out the fate of that plucky little spacecraft.

And so we say goodbye to *Beagle 2* and the hopes of a handful of British scientists who dared to dream as adults the visions they dreamed about when they read their science fiction as children. Their Spitfire may have crashed and burned on the plains of Isidis Planitia, but their spirit lives on.

## THIS NEW EDEN

It was one of the most moving things I have ever seen: two Martian warriors facing each other across the royal blue cloth covering the long table at the head of the Berrill Lecture Theater at my academic home, the Open University at Milton Keynes. On November 7, 2006, Colin Pillinger balanced on his crutches as he stood to welcome Steve Squyres to talk to the university about the staggering successes of the Mars Exploration Rovers, *Spirit* and *Opportunity*.

For Colin and all of us at the Open University, the occasion was tinged with melancholy because it was suffused with thinking about what might have been. What if *Beagle 2* had possessed sensors to image the terrain and state of the Martian atmosphere as she flashed across the Martian sky, as the Mars Exploration Rovers' landing pods had? What if her landing ellipse had been a kilometer to the east or west? Might she have missed that tiny crater's rim and not spiraled in, thereby losing radio contact with Earth? Might she have landed safely on the good, flat Martian soil of Isidis Planitia, her solar panels unfolding to catch the rays of the distant sun as her marvelous mechanisms swung into action? What if the carbon machine had produced the goods? What if Dennis Leigh's tiny mass spectrometer had produced definitive evidence for the presence of life on Mars? What if we now knew that we are not alone in the universe?

As we smarted from the loss of *Beagle 2,* Steve Squyres and his team pulled off one of the most remarkable feats of Martian exploration since Schiaparelli drew his first map of the surface. Mars Exploration Rover-A (*Spirit*) was launched on June 10, 2003, and her sister ship, Mars Exploration Rover-B (*Opportunity*), followed on July 7. *Spirit* landed just south of the Martian equator (175 degrees east and 14 degrees south) on January 4, 2004, while *Opportunity* landed in the region known as the Meridiani Planum (the Meridian Plateau), smack-bang on the Martian prime meridian (0 degrees east) and even closer to the Martian equator at only 5 degrees south on January 25, 2004.

The Viking missions had shown just how difficult it could be to select appropriate landing spots on Mars. Not only was it important to find a safe landing site that wouldn't wreck your billion-dollar spacecraft, but there was also the issue of what you would find once you set down. The Mars Exploration Rovers (MERs) were the direct descendants of the *Sojourner* rover of the Mars Pathfinder mission, which had tested the technology of roving vehicles on Mars in the late 1990s. The MERs were marvels of engineering—robotic geologists programmed to work for ninety days on the hostile surface of Mars under the guidance of distant mission control at JPL. They could move, image their surroundings in real time with stereo cameras at the same height as a pair of human eyes, pick up rocks, turn them over, drill into them, and analyze their chemistry. Their six wheels and massive solar panels gave them an extraordinary range across the surface of Mars, but that didn't mean they could go anywhere. There was no one to pick them up if they fell over, no one to dust off the solar panels if a careless piece of maneuvering allowed a rock to fall on them, and no auto shop to change a wheel if one should become damaged. As a result, extreme care went into the selection of the Mars Exploration Rovers' landing sites—from the point of view of safety *and* maximizing the science the mission would return.

Gusev Crater is about one hundred miles in diameter and was chosen as *Spirit*'s landing site because a sinuous channel, Ma'adim Vallis, runs into it from the south. The growing number of images from orbiting spacecraft such as *Mars Global Surveyor* and *Mars Odyssey* had gradually convinced scientists that there had once been liquid water on Mars, and it seemed very likely to the MER landing team that the Ma'adim Vallis channel had once contained some. The team consequently thought that Gusev Crater might once have been the site of a body of standing water—perhaps even an ancient crater lake. Gusev Crater thus seemed like a perfect candidate for a landing site, given that the primary objective of the missions was to search for evidence in the geological record for past conditions suitable for life—including water.

*Opportunity*'s landing site at Meridiani Planum was chosen because spectroscopic scans from orbit had indicated the presence of sizable quantities of the mineral hematite—an iron- and oxygen-containing mineral only known to form in the presence of water. It, too, was thus considered a prime candidate for a landing site.

Throughout the planning for the MER mission, the science team were remembering the lessons learned thirty years before when the Viking spacecraft had set down on Mars. Those landers had been trapped by both their immobility and the sheer ignorance—at the time—of the surface of Mars. Had they simply landed in the wrong place? Was that why neither had shown the slightest hint of any organic molecules on Mars? The mistakes made during the Viking missions would not be happening again. The Mars Exploration Rovers were not only mobile, but their landing sites were now informed by decades of imaging and chemical analysis from orbiting spacecraft. Science and safety would be the guides to their landing sites. What was more, neither of the MERs were equipped with experiments to test for the presence of life. They were robotic geologists tasked with finding out whether the conditions necessary for the presence of life had *ever existed* on the surface of Mars. Perhaps

America was anxious not to have another embarassing debacle over life experiments as had dogged the Vikings. This time there would be no public hand-wringing over whether the right experiments had been included or what the results might mean. The MER team was primarily geologists, not biologists.

It was this omission that incensed Pillinger. He once told journalists, who had asked him why he was bothering to search for life on Mars when Steve Squyres was going there with a billion dollars worth of equipment, that Squyres was *not* going to search for life; he was merely going "sightseeing." But Squyres did not take offense. Indeed, on that Christmas morning when *Beagle 2* went AWOL, Squyres had been the first to phone Pillinger and offer comfort.

When at last *Spirit* opened her eyes after the dust of landing settled, she saw a flat plain with a slightly rolling surface littered with small rocks. On the horizon was a range of small hills. Almost the first act the MER team performed after arriving was to give this strange new place a human feel by naming it: *Spirit*'s landing site became Columbia Memorial Station in memory of the astronauts who perished in the *Columbia* space shuttle disaster in 2003. On January 27, 2004, the thirty-seventh anniversary of the *Apollo 1* fire, MER scientists named the three small hills to the north of the landing site the Apollo 1 Hills and its three peaks after Gus Grisson, Ed White, and Roger Chaffee. The range of hills to the east of *Spirit* became the Columbia Hills, and seven of its peaks were named after *Columbia*'s crew.

The *Spirit* team had a brief scare when the spacecraft's onboard flash memory system started acting up. Demonstrating how far "on-the-fly" spacecraft repair has come, the MER engineers fixed the problem remotely. It turned out to be software related, and JPL scientists prepared a software patch that they uploaded to both spacecraft to prevent its reoccurrence.

*Spirit* started work as soon as she rolled off her landing pad. As predicted, Gusev Crater turned out to be an old crater lake. Some exposed outcrops nearby showed faint layering suggestive of sediment deposits. Near where Ma'adim Vallis enters the crater, topographic indications resembled earthly river deltas. Terrestrial river deltas can take tens or hundreds of thousands of years to form, and if the area around Ma'adim Vallis is a delta, water may have existed on Mars for a long period.

*Spirit* was sent to Mars to discover whether Gusev Crater is a fossilized lake bed. The scientists found out very quickly that the crater shows only a little evidence of the action of surface water. They estimate the water disappeared 3.5 billion years ago—at about the time that life started on Earth. They also discovered, however, that the region suffered intensely from volcanic activity.

*Spirit* had landed two and a half kilometers from the Columbia Hills, and they looked interesting. Yet the machine had only been designed to last ninety days and range about six hundred meters from the landing site. For three months, the team investigated a rock that they named Adirondack, discovering there olivine minerals that confirmed that, yes, there had once been a lake at Gusev but that its sediments had long ago been buried by lavas. A Martian day, or a Sol, is slightly longer than an Earth day, and when Sol 100 came around with *Spirit* still functioning perfectly, the team decided to trust the vehicle and set off for the Columbia Hills.

*Spirit* reached Husband Hill on Sol 156. Although the hill was small by Earth standards—only about the height of the Statue of Liberty—to a robot the size of a golf cart operating millions of miles from home whose instructions took almost ten minutes to reach it, the obstacle seemed impossibly high. Yet, what was the point of having had the luck of getting this far—almost twice as long as the mission lifetime and five times as far—if they didn't keep using the machine? The team decided to have *Spirit* climb Husband Hill. It was a good decision for another reason. *Spirit*'s first Martian winter was almost upon it, and the solar panels would soon be receiving much less energy. They would have to do something they had not expected—overwinter the craft and hope it would start up again the next Martian summer. They parked *Spirit* on a northfacing slope and waited and hoped.

It turned out that parking on a northfacing slope was inspired. The craft continued to receive enough power to operate and in the Martian spring was able to resume activity. Steve Squyres recently acknowledged that, without that northfacing slope, *Spirit* would not have survived that first Martian winter.

When *Spirit* resumed geological activity the next summer, the first thing she found was that the Columbia Hills were geologically complex. There was, for example, evidence for the presence of the mineral goethite—an oxygen-bearing mineral formed on Earth in the presence of water. In addition, glass formed by volcanic activity and granular rocks that looked as though they had been placed there by the action of volcanism were discovered. It was an extension and an amplification of what *Spirit* had found on the plain below. Both water *and* volcanism had formed this region of Mars billions of years ago.

Currently, *Spirit* is ending her second winter on Mars. She successfully scaled the summit of Husband Hill and has taken some of the most remarkable panoramic photographs in the entire solar system. Unfortunately, one of her front wheels has locked, and she is now traveling backward, dragging it behind her. She has been operational on Mars for over a thousand Sols, and nobody knows how much longer she has left. Every day is a bonus.

While *Spirit* was reconnoitering Gusev Crater, on the other side of Mars her sister ship had landed; team scientists commemorated the astronauts who died in the 1986 *Challenger* disaster, naming her landing site Challenger Memorial Station. A short while after landing, when *Opportunity* opened her eyes; what she saw was astounding. *Opportunity* had, purely by chance, landed in a twenty-meter-wide impact crater in the Meridiani Planum. The ancient asteroid that had plummeted into the Martian surface, creating the crater, had dug a serious chunk out of the surrounding rocks. As soon as the rover turned its eyes on the walls of the crater, the waiting scientists at JPL saw what they had come to Mars to see: the unmistakable layer-cake topology of sedimentary rock—rocks that had been deposited in water.

*Opportunity*'s landing crater was named Eagle in acknowledgment of the first manned spacecraft to land on the moon and a nod to the golfing expression "hole-in-one," which was exactly what *Opportunity*'s landing had been. *Opportunity* landed at Meridiani Planum to explore a chemical "beacon" that had been flashing on-and-off at orbiting spacecraft for several years, apparently marking the presence of hematite, a mineral that only forms on Earth in the presence of water. In Eagle Crater, *Opportunity* saw layered sedimentary rocks and sandstones containing masses of water-formed hematite in the form of concretions that the team named "blueberries."

After investigating Eagle Crater (and, once again, at around the critical ninety Sol time-horizon), *Opportunity* set off to the east for Endurance Crater, where she found three geological units that showed evidence of wind-blown sedimentation. The upper unit showed clear evidence of subaqueous stratification—proof positive that water had once flowed on the surface of Mars. It was exactly what *Opportunity* had come to Mars to find, and the excitement of the science team was palpable—Mars had indeed once been wet.

Yet the largest question of all remains unanswered—did Mars once harbor life? At the present time, the evidence seems stacked against it, for both rovers' data suggest that the surface of Mars is, and perhaps always was, oxygenating, acidic, and saline. Compare this situation with that on Earth at the time life got started. The atmosphere here then was reducing and with a neutral pH. The oceans were salty rather than saline. Today's Earth does host extraordinary single-celled organisms that can live in very acidic environments—the so-called acidiphiles. But, although they can live in concentrated sulphuric acid with a pH of 1, inside the cell the pH is a neutral 7, just as it was when their remotest ancestors evolved in the primordial seas of Earth. Acidiphiles have evolved sophisticated ion pumps across their cell membranes that protect them from the unbelievable harshness of their environment. Could they have originally evolved in such acidity? Probably not.

*Opportunity* has now probed further south and stands on the lip of Victoria Crater, by far the largest crater she has visited so far. It shows evidence of distinct stratigraphic layering, and the future looks fascinating. But, like *Spirit*, *Opportunity* has now passed her thousandth Martian day. One of the actuating coils in her robot arm has expired, and, although there is a redundant coil, the science team knows that it is one broken wire from massively reduced dexterity and two from paralysis. Time will tell.

*Spirit* and *Opportunity*'s landing sites were a testament to good mission planning and a certain element of luck. Both of these factors were augmented by peerless engineering. The two craft were designed to withstand the brutal conditions on the Martian surface for ninety days; as I write this in November 2006, both rovers have been operating for over a thousand days. Many of the rovers' findings would have remained locked in the mystery of the Martian soil if their good, US of A engineering had not allowed them to persist for so long.

Recently, the most extraordinary picture was sent to Earth by NASA's *Mars Reconnaissance Orbiter* (*MRO*), which in November 2006 pulled into the now rather crowded skies over Mars. *MRO*'s ultra high-resolution camera pointed its bionic eyes at Victoria Crater and saw tiny *Opportunity* sitting on the crater's rim, ready to resume its geological investigations when the harsh Martian winter ends. For the science team, it was an emotional moment: to see again the spacecraft last seen in their White Room at Cape Canaveral shortly before she left Earth and that they had never expected to see again.

*MRO* is the latest in a long line of orbiters that have beaten a now well-trodden path to Mars. Despite the loss of *Beagle 2*, the European Space Agency's *Mars Express* spacecraft has been sending back the most extraordinary stream of images since arriving in 2004. From these images, John Murray at the Open University has definitively shown that many of the red planet's surface features were formed by the action of running water. John has also been able to show that Mars *still* has water. In several places, ice fields lurk just beneath the planet's rusty surface. Discovering water sources on Mars brings us one step closer to sending human visitors—and then colonists—to the planet.

Finally, I must say something about what *Spirit* and *Opportunity* mean to practicing geologists like myself. The thought is not mine; it comes from my friend Nick Rogers, whose office is next to mine at the Open University and who sits on the council of the Geological Society of London, the oldest learned geological society in the world. Nick is a fascinating man to talk to, urbane, cultured, and a committed geoscientist. Having seen Steve Squyres's talk, he commented to me that he was amazed at the technology that the two robots were using, and all in the service of the most fundamental and oldest part of our particular science—field geology.

His point is well made. A hundred and fifty years ago, geologists like Charles Lyell, Roderick Murchison, and Charles Lapworth were roaming the hills of Britain establishing the geological framework of our country, a framework that was then used to interpret much of the rest of the world simply because the geology of Britain is among the most varied on Earth. Britain is a geologist's dictionary.

Those geological pioneers were armed with hammers, notebooks, and hand-lenses. Today, *Spirit* and *Opportunity* are doing the same thing on a world seen most often as a red dot in the sky. Their hammers may be robotic arms with automated rock abrasion tools, their notebooks may be flash-memory computer chips and high-gain antennae, and their hand-lenses may be high-resolution digital cameras, yet the principle remains the same. We have already spent billions to study the geology of Mars and do what we once did on Earth.

Every penny is well spent. The human spirit—scientific and inquisitive—is ready for the next step on our road to the stars.

# 6

# The Outward Urge: Asteroids

In 1776, seven years after the appearance of Halley's Comet, the German scholar Johann Titius formulated an idea that would profoundly affect planetary studies. He showed that the distance of the planets from the sun could be described mathematically, and that the actual position of the planets fit his calculations almost perfectly. It was an incredibly important finding because it confirmed Newton's influential theories on a logical, mathematical structure underpinning the solar system and the universe. Titius's theory was refined and confirmed a few years later by another German astronomer, Johann Bode (who used it without crediting Titius), and the idea became enshrined as the still-controversial Titius-Bode law.

Soon, though, the Titius-Bode law appeared to be seriously flawed because it predicted the presence of a planet 2.8 astronomical units (AU) from the sun (the Earth is at 1 AU, Mars is at 1.5AU, and Jupiter is at 5.2 AU), and there was no sign of such a planet. But the

Titius-Bode law could not just be discarded, especially following the discovery of Uranus by William Herschel in 1781, which, at 19 AU, was exactly where the Titius-Bode law predicted it would be. An informal group of astronomers, led by the Hungarian Baron Franz von Zach and calling themselves the "Celestial Police," set out to prove the validity of the Titius-Bode law by finding the missing planet.

It seemed that success was near on January 1, 1801, when the Italian astronomer Giuseppe Piazzi, director of the Palermo Observatory in Sicily, stumbled across a bright speck in his telescope. At first he thought it was an eighth-magnitude (that is, very dim) star. Subsequent observations on successive nights, however, showed that the "star" was moving at a significant rate across the star-field. This bright speck could only be a planet, or something very like it. Piazzi, with the help of Bode, calculated the orbit of the strange object, and they found that it was broadly circular, like that of the other planets. The brilliant young German mathematician Carl Gauss refined their calculations and predicted when and where the object would appear again. On New Year's Eve 1801, his calculations were triumphantly proven when the object appeared exactly where he had said it would be. The precision of Gauss's orbital calculations allowed him to estimate its distance from the sun: 2.8 AU. It seemed that Piazzi had found the missing planet predicted by the Titius-Bode law, and he named it Ceres.

Within months, questions began to pop up about the new planet. Its orbital inclination (its angle to the plane of the ecliptic) was very high, it was much smaller than any of the other planets. Also, in March of 1802, the astronomer Heinrich Olbers found a second object traveling in exactly the same orbit, something that had never been observed in the history of planetary astronomy. Olbers named this object Pallas. Its orbital inclination was even greater than Ceres—a massive thirty-four degrees—which made it more like a comet than a planet. Since they had no tail or nebulosity, however, they quite clearly were not comets, and William Herschel suggested that they be named "asteroids."

For a long time one of the most important questions in planetary astronomy was whether asteroids were the remains of a planet that had exploded or collided with another long ago, or fragments of the primordial planetary disk that had failed to coalesce. The exploded planet theory dates back to 1802, only a year after Ceres was found by Piazzi and the year that Olbers discovered Pallas. Olbers was born near Bremen in northern Germany and, like many scientists of the time, practiced medicine. At night he was a dedicated and enthusiastic amateur astronomer, who had converted the upper floors of his home into an observatory. It was here that Olbers identified and named Pallas. Five years later, his assiduous observations led him to discover another asteroid that, in a generous gesture, he allowed his friend Carl Freiderich Gauss to name. Gauss named it Vesta.

The problem for nineteenth-century astronomy was that these three asteroids all lie in the same orbit—precisely where the Titius-Bode law predicted the fifth planet from the sun should be. Olbers proposed that the asteroids—which at the time were still called "planets"—were the remnants of a larger planet that had inhabited that region of space and been destroyed by some unknown cataclysm. On the basis of this theory, he declared that still more planetary fragments would be found, a prediction that was vindicated when Karl Harding discovered Juno on September 1, 1804. Not until a hundred years later was Olbers's hypothetical exploded planet dignified with a name. The Russian meteorite specialist Yevgeny Krinov suggested calling it Phaeton after the son of Helios, the sun god in Greek mythology.

Today the Phaeton theory is largely discredited by astronomers because we have a surer understanding of the nature of gravity, and of Jupiter's gravity in particular. Jupiter is 318 times more massive than the Earth and is the second most massive object in the solar system after the sun. The theory now is that there *should* have been a planet in the position predicted by the Titius-Bode law but that it never had a chance to form because its components have been permanently disrupted by the massive gravity of its neighbor. The intense scientific interest in the asteroids is thus understandable: the clear implication is that they are composed of material that has never been formed into a planet. The asteroids are the unused building blocks left over from the formation of the solar system.

## THE GRAVITY SLAVES

To understand a little about the strange force of gravity—so ubiquitous that we don't think about it in our day-to-day lives but so important that it can prevent the formation of a planet—we need to briefly visit a small market town in the English fens during the mid-seventeenth century. Isaac Newton was accepted into Trinity College, Cambridge, in 1661. Cambridge was the intellectual and scientific hothouse of the age, and the young Newton was quickly exposed to the new theories of Copernicus, Kepler, and Galileo as well as those of Descartes, Gassendi (who had recently formulated the atomic theory of matter), Robert Boyle, and Henry More. These thinkers would vitally influence the formulation of Newton's ideas about gravity and orbital mechanics.

Like Galileo before him, Newton's life was not without its conflicts. The scientist Robert Hooke, for example, delighted in contradicting him. In 1679 Hooke and Newton were discussing the path of a freely falling object. Newton proposed the path would be a spiral, falling toward the center of the Earth; Hooke countered that it would be an ellipse. From this discussion, Newton started formulating the mathematics of orbiting bodies.

From autumn 1684 through spring 1686, Newton expanded and refined his calculations, finally writing them up in a work that was to become one of the most important physics books of all time, *Prin-*

*cipia Mathematica.* Newton knew that he could explain the movement of objects on Earth with his new laws, and he began to wonder if these laws could explain the motion of the planets as described by Kepler's laws, the most important of which is that each planet orbits the sun in an ellipse with the sun at one focus. To do so he would need a force to act on the planets, and he realized that such a force existed: gravitation. By combining two of his greatest intellectual achievements, the laws of motion and the law of gravitational force, Newton was able to explain why the planets moved as they did in accordance with Kepler's laws. Their elliptical orbits are explained by applying Newton's laws of motion. His new theory actually improved upon Kepler's laws.

It was a staggering achievement. For two thousand years, since the time of Aristotle, humans had wondered what held the stars and the planets in their courses. Now they knew.

## THE GALAXY RANGERS

So asteroids of the solar system's inner asteroid belt are small, rocky bodies that orbit the sun between Mars and Jupiter but that cannot coalesce into a single body because of the gravitational forces exerted by their massive neighbor, Jupiter. They range in size from Ceres, about a thousand kilometers across, down to pebbles. To clear up the tidal wave of nomenclatural confusion that surrounds asteroids, in August 2006 the International Astronomical Union (IAU) classified all objects smaller than Ceres as Small Solar System Bodies. This category naturally includes the majority of asteroids, most trans-Neptunian objects (objects that orbit the sun outside the orbit of Neptune), and comets. Small Solar System Bodies (SSSBs) are smaller than dwarf planets and do not exhibit hydrostatic equilibrium, which is a fancy way of saying that they are not round.

It is not yet clear whether the IAU will impose a lower boundary on SSSBs or if the term will encompass all material down to the level of meteoroids. A meteoroid is a relatively small object (boulder to sand-grain sized). Meteroids can be seen leaving fiery trails in the upper atmosphere when they streak through it as meteors. If any material survives and hits the ground, it is referred to as a meteorite. Thus *meteoroid*, *meteor*, and *meteorite* are terms that describe the same object at different stages and localities of its existence.

The Phaeton theory began to be questioned as more and more data about the nature and size of the asteroid belt accumulated during the twentieth century; it became clear that there was not enough mass in the asteroid belt to make a small moon, let alone a planet. George Wetherill—the man whose ideas about planetary impacts in the 1970s and 1980s helped settle the debate over the moon's formation once and for all—also settled the debate over the formation of the asteroid belt. In 1960, after a first career at the Carnegie Institution's Department of Terrestrial Magnetism in Washington, DC, Wetherill moved to California where, for the next fifteen years, he would work at UCLA. Wetherill was the kind of scientist interested in big questions with big answers. For example, while at Carnegie he had sorted out some of the biggest problems in the science of geochronology—the absolute dating of rocks using radioactivity. At UCLA, he changed tack and undertook an even bigger challenge—understanding the forces that control the orbits of asteroids and comets and estimating the chances that they might hit Earth.

At UCLA, Wetherill precisely calculated the chances of a body orbiting in the asteroid belt between Mars and Jupiter colliding with the Earth and showed that most meteorites found on Earth must have originated in the asteroid belt. It was Wetherill, too, who showed that it was possible for asteroidal meteorites hitting Mars to explode Martian material into orbit—material that in theory could find its way to Earth. Wetherhill's predictions have been spectacularly verified by the subsequent finding of dozens of meteorites from the red planet. His work also highlighted the importance of geochemical measurements on meteorites, a technique that, as we saw in the last chapter, became important in analyzing the Martian meteorite ALH 84001, which in 1996 NASA researchers suggested contained fossil evidence of Martian life.

After his seminal work on the origins of the moon and the terrestrial planets—conducted during the final phase of his career back at Carnegie—Wetherill retired in 1991 but, like so many brilliant scientists, continued to do research. He returned to his earlier love of asteroids, perplexed by the question of the asteroid belt's missing mass. Since there was not enough material there to have made a planet, clearly a planet could not have been destroyed there. This conclusion left only two choices: either something had prevented the formation of a planet in the first place, or something had vacuumed up much of the extra material. In either case, there was only one suspect: Jupiter and its giant gravity field. Wetherill realized, however, that no one knew the exact mechanism by which Jupiter might have precluded the formation of mythical planet Phaeton.

Deep-space measurements by the probes sent to the asteroid belt over the past fifteen years have shown that Jupiter's gravitational influence only extends to a part of the asteroid belt. In 2001, just five years before he died, Wetherill, working with his colleague John Chambers, showed that there may once have been enough material to make several Earth-sized planets in the asteroid belt but that they were precluded from forming by a combination of complex gravitational forces and random events. This assertion was extraordinary because, until that time, the asteroid belt had been thought to be the exclusive domain of small planets and planetismals. The planetesimal hypothesis of planet formation is nothing more than the twenty-first-century version of Laplace's nebular hypothesis, which, as we have seen, states that the planets formed when the smaller bodies of the primordial solar nebula collided and stuck together. There were two distinct regions of this solar nebula—the inner region where nonvolatile material survived to form the rocky Earth-like planets, and an outer region where temperatures were sufficiently low to allow hydrogen-bearing compounds to solidify into the gas giants. The masses of the planets can all be explained by this theory—with the notable exception of the asteroid belt.

Chambers and Wetherill theorized that the planets began to form across the primordial disk relatively rapidly—certainly in less than a million years. But the proto-planet forming in the region of the asteroid belt was repeatedly disturbed by Jupiter's rapid growth. The asteroid proto-planet's distance from the sun was also critical, because it orbited the sun two times for each single orbit of Jupiter. This 2:1 resonance meant that it experienced regular shoves from Jupiter, gradually deforming its orbit. Its irregular orbit greatly increased the risk of a collision with another of the rapidly forming proto-planets, putting it in danger of being ejected from the solar system or flung into the sun.

As Jupiter grew, its orbital resonances became stronger but were confined to discrete zones within the asteroid belt. Collisions between proto-planetary fragments were common, and any debris that strayed into one of the danger zones was quickly absorbed by Jupiter's colossal gravity field. Eventually only two or three planet-sized fragments remained.

Chambers and Wetherill calculated that 90 percent of the asteroid belt's material was cleared away in only a few hundred million years—a tiny fraction of time when scaled against the age of the solar system. In addition, their computer modeling experiments showed that, if they elongated the orbit of Jupiter (or Saturn) by even modest amounts, the time needed to sweep the asteroid belt clean of its excess mass shortened to only a few million years. They found that the resonance "danger" zones became wider and stronger if the orbits of the perturbing planets were elongated. It was clear that, from an early stage, Phaeton had no chance of survival.

Ironically, there was a feedback effect on the orbits of the gas giants, particularly Jupiter. All the gravitational finagling depleted the gas giants' stores of orbital energy, and as a result, their orbits became less elliptical and more circular. This change progressively reduced the gravitational perturbation of the remaining asteroid belt proto-planets.

Chambers and Wetherill's theories also suggest a mechanism for the origin of the Earth's oceans. So great were the perturbations caused by these early orbital resonances that they may have sent comets—which are mostly composed of ice—flying through the primordial solar system into the Earth, explaining how so much water came to our planet so early in its history. Since comets also contain the chemical precursors of amino and nucleic acids, they may also have supplied the raw material that started evolution on Earth.

## BELTED BY ASTEROIDS

The down side to living in an active solar system is the same one we get from living beside a busy road: we have to look out for moving objects. The importance of that vigilance is reinforced every time we look at the moon. Its benign face is pockmarked by the scars of 4.5 billion years of close encounters with impacting objects. Its countless millions of craters are so densely packed that they overlap each other. Until formulating the theory of plate tectonics, however, we had no reason to suspect that our own planet must have suffered a similar outrage of bombardments but was able to cover up its scars.

The University Museum at Oxford is an extraordinary building. It was built in the nineteenth century as a shrine to the new science of natural history and, very soon after its completion, was host to one of the most fateful meetings in the history of science—when Thomas Henry Huxley, Darwin's bulldog, met Bishop Samuel Wilberforce in a famous debate over the theory of evolution. It was also the building where I was housed when I arrived in Oxford during the summer of 1988 as a greenhorn postdoctoral researcher. Almost immediately I became aware that the university that year was playing host to a remarkable conference—the annual meeting of the British Association for the Advancement of Science (BAAS).

It is incredible to me now that I had no inkling of this meeting before I arrived in Oxford, for the BAAS is one of the most important meetings on the British scientific calendar. I well remember going into the University Museum's lecture theater, just a few yards from my office, and finding two eminent scientists locked in a debate over the interpretation of the fossil record—not quite as momentous as the debate between Huxley and Wilberforce, but historic in its own right. One of the scientists was the dilettante of British paleontology, Bev Halstead, and the other was Walter Alvarez, the son of the formidable physicist Luis Alvarez.

The argument that sunny day in Oxford was whether or not the dinosaurs had been killed off instantaneously by the arrival of an asteroid or whether they had died out gradually over several million years at the end of the Cretaceous period. Halstead was a proponent of the gradual extinction theory, and I can remember the diagram he so painstakingly drew on the lecture room's blackboard. It was a diversity diagram of the style favored by paleontologists throughout much of the twentieth century. Looking like the side view of a turnip, it showed a bulge at the beginning of the Cretaceous, when dinosaur species numbered in the thousands, and narrowed toward the top as those species became progressively fewer. It was always a strange way to draw a diagram—a single line on a conventional graph with x and y axes would have more than sufficed—and someone in the audience asked Halstead why he drew it in that peculiar way. The answer—which Halstead did not condescend to give—is that paleontologists had always drawn their diversity diagrams in this cartoon style, perhaps to hide the paucity of their dataset and to reflect the preconception then underlying their science, namely, that evolution and extinction happen gradually over millions of years. It was an idea that was in the process of being spectacularly overturned by Alvarez's theory. And that spectacular overturn has its origins in one of my favorite parts of the world, a place where medieval hill towns drowse in the lambent sunshine of southern European summers, the air thick with the scent of juniper and the long afternoons ticking quietly by, marked only by the clicking of crickets.

The importance of the Bottaccione Gorge to a geologist is that it has an almost uninterrupted sequence of limestone strata tilted on end so that it is possible to walk along its length, therefore walking through time between the middle Cretaceous and the middle Tertiary, a span of almost 100 million years. As a result, the gorge has been intensively studied by geologists. They have dissected its limestone ribs to reveal the evolving succession of tiny microfossils along its length—the remains of plankton that lived in the surface layers of the ocean long before the reign of man—they have measured the oscillating direction of magnetism held in its rocks, and they have analyzed the changing chemistry of the world's oceans among its tilted strata. About halfway along its length is a thin layer of clay that slopes at an angle of about fifty degrees from the road to the top of the cliff. Today this layer of clay has been so heavily excavated by generations of geologists that a deep gash has been carved into the face of the cliff.

The reason for this geovandalism can be traced directly to the Alvarez family's efforts, for the Bottaccione Gorge is where Walter and Luis Alvarez changed the face of paleontology. That clay layer is the physical manifestation of the boundary between the Cretaceous and the Tertiary periods (known as the K/T boundary) of Earth history (or, more correctly, the Mesozoic and Cenozoic eras), a major extinction horizon when the dinosaurs, and many other groups of organisms, became extinct. That specific time represents the starting gun for our own evolution, for mammals rapidly filled the ecological niches vacated by the dinosaurs.

By the 1970s, Walter Alvarez and his colleague Bill Lowrie had been working in the Bottaccione Gorge for several years. They were interested in calibrating the rate at which the gorge's limestone sediments, and especially the clay layer at the K/T boundary, were deposited. It is relatively easy to study the passage of time in limestone because the pattern of magnetic change in the rocks can be compared to a standard pattern assembled from deep-sea cores. But in this instance, the clay layer is too narrow (only one or two centimeters thick) for this approach to work.

Walter presented the problem to his Nobel Laureate father who, being a physicist, rapidly and unsurprisingly arrived at a physical solution to the problem. Luis suggested that they use the concentration of rare elements in the clay layer to estimate the passage of time. Alvarez and Lowrie decided to analyze the iridium content because it is a scarce element at the Earth's surface; it is only found deep in the Earth's interior and in outer space. They knew measurements were going to be difficult because iridium is only present on Earth in the parts per billion range. They would need to employ a relatively new technique, neutron activation analysis, to find conclusive data, so they sought out Frank Asaro, the local expert in the technique at Berkeley and a man with access to the necessary instrumentation, which is to say, a nuclear reactor.

The measurements were fraught with technical difficulties, but they eventually came through in the early summer of 1978. The data were not what they expected; instead of finding amounts of iridium in the hundredths of parts per billion, they found it in three parts per billion—three hundred times more than what they had expected. The team scrutinized the experimental technique and could find no fault (despite having to correct for the iridium in an analyst's wedding ring that had accidentally contaminated the first batch of results). Eventually they were forced to conclude that the measurements were correct. To be certain of their conclusions, they would need to find another K/T boundary section and make measurements to corroborate the first findings.

Walter thus went to the eastern coast of Zealand in Denmark to an area known as Stevns Klint where the K/T clay layer lies embedded halfway up a cliff of soft chalk and the waves of the Baltic Sea lap gently at its foot. The measurements from this new suite of samples showed the same anomalously high levels of iridium. It was the proof the scientists needed to be certain of an astounding conclusion: since such high levels of iridium could only come from outer space, the Bottaccione Gorge clay sample and the Zealand cliff chalk sample proved that an extraterrestrial agent was responsible for the demise of the dinosaurs.

Asteroids, as we have seen, are the unused building blocks from the formation of the solar system and so contain high levels of iridium. Luis reasoned that an impacting asteroid, if of sufficient size, would not only provide the iridium signature but would throw up enough dust to completely obscure sunlight and shut down photosynthesis on the planet. Similar effects on a smaller scale were observed after the 1883 eruption of Krakatoa in Indonesia; the volcano's massive eruption threw up enough dust to spread across the globe. The effects of an asteroid's impact would sever the food chain at its base on land and in the ocean and would rapidly lead to mass extinctions among many different animal and plant groups.

The asteroid impact hypothesis became front-page news around the world and marked an event in the history of paleontology every bit as profound as the K/T boundary itself was to the history of life. After Walter and Luis Alvarez published their theory in *Science* in 1981, paleontology took on a new role as a predictive science; paleontology's dalliance with comparative planetology transformed it. Over the past two decades, the idea of impacting asteroids has gone from rogue scientific theory to general acceptance in the scientific community as well as becoming part of popular culture's knowledge set. One thing is certain about impacting asteroids: they will hit the Earth again—it is simply a matter of when. Accordingly, there is intense interest in calculating the orbits of Earth-crossing asteroids, which has done much to rekindle interest in the asteroid belt.

About a thousand near-Earth asteroids—or objects that intersect the Earth's orbit—are known, varying in size from barely larger than a stone to some over thirty kilometers in diameter. Near-Earth asteroids ultimately either fall into the sun or fly into a planet, with an average lifespan of a comparatively short—to a geologist—100 million years. The pool of potential planet killers is constantly renewed by objects from the asteroid belt that are nudged out of their existing orbits by random gravitational perturbations from other asteroids and by mighty Jupiter.

Apart from the Alvarez hypothesis, two other events of the past thirty years brought the threat of near-Earth asteroids to the forefront. On March 23, 1989, the 300-meter diameter asteroid Asclepius missed Earth by only 400,000 miles. If it had arrived only six hours earlier, it would have collided directly with our planet, causing the largest explosion in recorded history. Five years later in July 1994, the world watched in awe as NASA's *Galileo* probe beamed back pictures of comet Shoemaker-Levi 9 colliding with Jupiter. Suddenly everyone—from paleontologists to the world's top military brass—was worried about asteroids. In less than fifteen years, extraterrestrial impacts had gone from being a theoretical problem of interest only in regard to the death of the dinosaurs to being a problem that would involve the world's military.

Scientists quickly started calculating the odds of collision, and the results were not reassuring. Every million years we can expect to be hit by several asteroids with an average diameter of one kilometer. Impact with a five-kilometer asteroid averages once every 10 million years. Some asteroids were already emerging as threats. Dramatically, on March 18, 2004, a thirty-meter asteroid passed less than 27,000 miles from Earth—less than one-tenth the distance to the moon. It was the closest near-miss ever recorded, and calculations show that similar-size asteroids probably come as close every couple of years.

Asteroid 1950 DA was first catalogued in 1950 but was then lost for fifty years. When it was rediscovered in 2000, its orbit was calculated to intersect with the Earth on March 16, 2880. That year may seem a long way off, but we are very far from finding a way to avert the apocalyptic disasters asteroid impacts would cause.

As a result, several projects have been initiated to study the possibilities and results of a collision. Perhaps the best known is LINEAR (Lincoln Near-Earth Asteroid Research), which began in 1996. LINEAR is a cooperative project between the U.S. Air Force, NASA, and MIT's Lincoln Laboratory for discovering and tracking near-Earth asteroids. LINEAR's eyes are the Ground-Based Electro-Optical Deep Space Surveillance Telescope based at the White Sands missile range in New Mexico. By the end of 2000, LINEAR had detected well over 200,000 new objects, of which more than 1,500 are near-Earth asteroids. A different project called Spacewatch, run by the University of Arizona, uses the ninety-cm Kitt Peak telescope in the southwestern United States. About twenty-five new near-Earth asteroids are discovered by Spacewatch per year. Spaceguard, another tracking program, is a loose agglomeration of affiliated international committees such as American Near-Earth Asteroid Tracking, the Italian Campo Imperatire Near-Earth Objects Survey (CINEOS), and the Japanese Spaceguard Association. The proliferation of all these organizations stems almost directly from Alvarez's discovery of the iridium anomaly in the Bottaccione Gorge nearly thirty years ago.

Many suggestions have been made on how to deal with the hazard posed by rogue asteroids. The impact of even a relatively common one-km object would be a disaster unparalleled in human history. A frequently proposed solution is to explode a nuclear device on an incoming asteroid. The problem with this approach is that the explosion might merely break the object into pieces, transforming it from one large impacting body into several small ones—without diminishing the danger. Others have suggested mounting mass drivers on the asteroid to push it off course. A more fantastical idea is to wrap an asteroid in aluminum foil, creating a solar sail to harness the sun's photons and redirect the asteroid's trajectory (similar to *Mariner 10* during its epic voyage to Mercury). Certainly the most innovative suggestion employs the Yarkovsky effect (Yarkovsky was a Russian engineer living in the latter half of the nineteenth century). The Yarkovsky effect results when more heat energy is radiated by the sun-facing (warmer) side of an extraterrestrial body than the darker (cooler) side; the emission of this greater heat energy exerts a (admittedly small) force on the body. Coating an asteroid with dark material on one side would preferentially warm it up by the absorption of solar radiation, which could then be used as an in situ motor to alter an incoming asteroid's trajectory.

Such schemes verge on fantasy and certainly would require a great deal of time to implement. The implication is that we will need decades if not centuries to prepare for the threat of an oncoming asteroid. Clearly much more research needs to be done—now a well-established goal of several space agencies. Only a hundred years ago, asteroids were so little regarded by the astronomical community that they were known as the "Vermin of the Skies." The survival of our species and perhaps even of our planet depends on increasing research and respect for asteroids; if they are vermin, they are among the most deadly we know.

## TURNING JAPANESE

One of the most exciting developments in space exploration during the last fifteen years has been the emergence of new space agencies in addition to the Americans and Russians. We have already seen how the European Space Agency has made major contributions to the study of Mars and Venus, and now the Japanese may be set to pull off a major coup. The Japan Aerospace Exploration Agency (JAEA) has landed a small spacecraft named *Hayabusa* on the surface of the asteroid Itokawa, where it is thought to have successfully taken samples. It is now headed back to Earth. If the sampling was successful, *Hayabusa*'s arrival home will be the first return of samples from an extraterrestrial body since *Apollo 17* splashed down in 1972.

The Japanese effort explicitly acknowledges the importance of asteroids to our understanding of the early history of the solar system. *Hayabusa* will provide a record of conditions just after the formation of the planets. A single sample from Itokawa could provide clues about the raw materials of which the planets are composed and provide a glimpse into the interior of the solar nebula 4.5 billion years ago.

*Hayabusa* is only the second mission dedicated to asteroid science. The first was the NEAR (Near Earth Asteroid Rendezvous) mission built by Johns Hopkins University, which arrived at and landed on (despite not being designed as a lander) the asteroid Eros in 2001. The NEAR mission was stimulated by the successful asteroid flyby of the *Galileo* spacecraft en route to Jupiter in 1992. *Galileo* showed that asteroid rendezvous is possible. NEAR was the first mission of NASA's much-hyped Discovery program, designed around the concept of better, faster, cheaper missions. The NEAR craft was built by the Johns Hopkins team in just twenty-six months. NEAR performed the first up-close-and-personal study of an asteroid. It was the first time that seismic profiling was performed on an extraterrestrial body of such diminutive size, and it showed that Eros was composed of several blocks of space debris loosely glued together. Incidentally, this finding was good news for the asteroid-diversion-by-nuke lobby, because it suggests that asteroids such as Eros will shatter like glass.

*Hayabusa*'s mission to return samples to Earth is much more ambitious. Launched on May 9, 2003, the tiny spacecraft used a novel ion-drive engine to power its way through space for twenty-eight months. Ionizing gas produces relatively low thrust, but it can be maintained for very long periods of time. Such motors may pave the way for future unmanned and possibly crewed missions.

The spacecraft also carried a new type of laser-ranging device that allowed it to self-correct its course during the landing phase of the mission. This, too, is an amazing achievement because, if a spacecraft can make its own landing decisions in real time, landings on complex surfaces will become easier in the future. *Hayabusa* arrived in September 2005 and took some astounding asteroid photographs. Incredibly, even the spacecraft's shadow on the asteroid appears in some of the photos as it prepares for landing.

Despite these achievements, problems arose after *Hayabusa* arrived at Itokawa. After surveying its target, the spacecraft approached for a series of encounters that would allow its sampling device to collect samples. Metal pellets were to be fired into the surface of the asteroid, kicking up material for collection. But due to technical problems, it remains uncertain that any material was collected. Data from the onboard computers during November 2005 imply that the projectiles were not fired. Materials may have made it onto the sample container by chance, however, and since little material is required, this part of the mission may still succeed—assuming the spacecraft can return home. At the time of this writing, this return may be problematic because further setbacks hit when, in November 2005, a thruster leak sent *Hayabusa* flailing wildly out of control. When contact was re-established in January 2006, the mission controllers realized that all the spacecraft's maneuvering fuel—required to align the ion-drives—was gone. *Hayabusa*'s engineers still have four functioning ion-drive engines, and the scientists remain optimistic. They're not out of the game yet; if the craft can be persuaded to return home, it will land at the old British air force base in Woomera, Australia, in June 2010. Go *Hayabusa*.

We have come a long way from the predicted existence of Phaeton at the start of the nineteenth century. But what is the current status of the Titius-Bode law? As mentioned at the beginning of the chapter, its predictive power is good, even though it did not predict the presence of Neptune—its most conspicuous failure—or of Pluto. But Pluto, as we shall see, is not a planet—it is a displaced Kuiper Belt object, and its current location in space should not be expected to be predicted by the Titius-Bode law.

Quite recently two French scientists came up with an explanation for the apparent predictive power of the law. François Graner and Berengere Dubrulle contend that the Titius-Bode law is a natural consequence of two properties common to any planetary system. Our current models of planet formation nearly always include as an important variable a cloud of material influenced by the gravity of the local star. Such clouds have two types of symmetry—rotational symmetry and scale invariance. Rotational symmetry is easy to understand. It simply means that, no matter how something—be it a primordial cloud of nebular material or a doughnut—is turned, it will always have the same symmetry. Scale invariance is the same kind of symmetry we get from looking at an ordinary coastline on Earth. At all scales it looks jagged. The two Frenchmen point out that scale invariance is important in many of the constraints in how planets form, and they have shown mathematically that, when the rotational and scale invariance of one of these constraints reaches either a maximum or minimum, the location of planets will always follow a simple relationship. In this relationship, the position of point "n" is proportional to "K" raised to the power of "n." If this kind of thinking makes you want to go around to the pub for a stiff drink (which is the effect it has on me), it can be summarized succinctly: The Titius-Bode law is a mathematical rule stating that the spacing of the planets from our sun ("n") is based on doubling each successive term ($K = 2$). Graner and Dubrulle show, in fact, that a better fit for the Titius-Bode law would be obtained by making $K = 1.7$, then raising it to the first power for Mercury, the second power for Venus, and so on, multiplying each result by 0.23. For the Titius-Bode law to work, however, not only must the primordial materials in the early planetary system have been both rotationally and scale invariant, so, too, must other conditions like temperature and chemical distribution. At the present time, the jury is still out on that one.

Finally, there is one curious—and perhaps heretical—thought. Given that the Titius-Bode law did not identify Pluto, and we now know on the basis of other evidence that Pluto is not a planet, what, then, of Neptune? Is that far guardian perhaps also a Kuiper Belt object? We must be careful here, for the gates of circular reasoning yawn, but the possibility cannot be discounted that the Titius-Bode law works and that the exceptions are not exceptions, because they are not, after all, true planets. A comforting thought is that, if there is a law to the distribution of planets, we may well find it when we look at other planetary systems in sufficient detail. Just as the Fibonacci series (a mathematical sequence that underlies the shape of snails' shells and the ordering of shoots on plants as well as much else) provides a semblance of order to our world, it is nice to think that out there somewhere there is a little order in the cosmos, too.

# 7

# The Eye of the Universe: Jupiter

In an office not far from my own at Open University sits a copy of an artifact no human will ever see again. The two originals, each securely bolted to the side of a *Voyager* spacecraft and protected by aluminum sheaths, resemble old-fashioned phonograph records but are made of pure gold. They were sent to the far reaches of our solar system and beyond to act as our calling card if either should ever encounter an advanced extraterrestrial race. The records were designed by a committee led by none other than Carl Sagan. Each contains an eclectic selection of audio and video recordings, from Johann Sebastian Bach to Chuck Berry, from the chant of New Guinea tribesmen to the roar of a Saturn V rocket blasting off its launch pad. On the records' protective sheath are images, meticulously designed, to communicate how to play them. But how do you communicate to an unknown intelligence? For Sagan and his team, it was a significant problem. They settled on writing the instructions in the universal language of mathematics—binary code—and basic diagrams.

At the ten o'clock position on the sheath is a diagram of the record and a representation of where to place the stylus and cartridge included with it. The speed at which the record must revolve—3.6 seconds for one revolution—is explained in binary arithmetic. The rotation information is expressed as multiples of the time period associated with a fundamental constant of the universe: the transition time of the hydrogen atom. Other images on the record show graphically how to play it and indicate that it contains both video and audio information.

One of the cleverest innovations on the record is its wristwatch: a two-cm-diameter area impregnated with a tiny dose of pure uranium–238. Every 4.5 billion years, half of that microscopic dose of uranium–238 will decay into lead. To determine how long the spacecraft has been journeying through the cosmos, an investigating extraterrestrial must simply measure the ratio of uranium–238 to lead–206.

Finally, and perhaps most poignantly, at the bottom left of each record cover is an image showing the location of our solar system with respect to fourteen pulsars—stars that emit radio waves at a precise frequency. Any intelligence with a knowledge of the interstellar geography of our corner of the universe will be able to locate Earth—the home of the small, hopeful civilization that sent the *Voyager* emissaries to the stars.

Since their launch in the late 1970s, the *Voyager*s have been speeding away from Earth at a velocity that far exceeds that of any other human-made object—over 38,000 mph. *Voyager 1* is already beyond the heliopause—the edge of the solar system where the solar wind stops—with *Voyager 2* not far behind. The golden records they carry are humanity's big "hello" to an indifferent universe, a message from the first flush of optimism that accompanied the dawning of our species' space age.

## THE STARMAN OF PADUA REVISITED

The *Voyager* spacecraft were not sent primarily to bring greetings to aliens. Their function was far more prosaic and has its roots in a cold January night at the beginning of the seventeenth century in Galileo's chambers.

Although Galileo's 1609 observations of the moon were revolutionary, their importance was eclipsed by other observations he made during the same period. For example, using his new telescope he proved that the planets were observable disks whereas the stars were not, implying that the latter are immeasurably farther away. But to many, Galileo's discovery of four satellites around Jupiter is perhaps his greatest astronomical contribution. The satellites were all discovered within the space of a few days during January 1610. On January 7, Galileo saw three objects near Jupiter, strung in a line like pearls. When he observed them again the next evening, they appeared to have moved. During the following week, he continued to observe the strange objects and made accurate drawings of their positions; by January 11 he had found a fourth object. All four objects (or "stars" as Galileo inaccurately called them) stayed close to giant Jupiter and appeared to be carried along with it through the sky, although they continuously changed their positions relative to each other and to Jupiter. After several weeks of observation, Galileo realized he was not looking at stars at all but at bodies orbiting Jupiter. He called them the "Medicean planets" in honor of his patron and numbered them I through IV. It was not until 1614 that the astronomer Simon Marius, based on a suggestion by Kepler, gave these moons the names by which they are known today: Io, Europa, Callisto, and Ganymede. These names were not officially adopted until the mid-eighteenth century when it became clear that there were many more moons orbiting Jupiter, making Galileo's numbering system too cumbersome.

Galileo's discovery of Jupiter's moons jolted astronomy to the same degree as Copernicus's findings. Now not only were there stars beyond our sun, even the planets of our solar system were not unitary bodies; they had their own orbiting minions—moons. It was the first hint that the universe is what we today call "fractal": similar to itself on a variety of scales.

In 1610 Galileo published his observations in a book titled *Sidereus Nuncius* (Starry Messenger). The book attracted immense criticism across Europe from those who doubted Galileo's observations, but Johannes Kepler, the young German mathematician who had shown that planetary orbits were ellipses rather than circles, came to Galileo's defense. Kepler wrote Galileo a strong supporting letter (which he subsequently published as "Dissertatio cum Nuncio Sidereo"—Conversation with the Sidereal Messenger) and confirmed Galileo's observations with his own, made with a recently acquired telescope. Kepler published his observations of the Jovian satellites as *Narratio de Observatis Quatuor Jovis Satellitibus* (Narration about Four Satellites of Jupiter Observed). His writings gave enormous support to Galileo, but that support was not reciprocated. In July 1610, when Galileo was sure that he had discovered two moons orbiting Saturn, he made no mention of them to his young admirer.

Despite Kepler's corroboration of his findings, the author of *Sidereus Nuncius* was soon in trouble. In 1616 the Pope asked the Inquisition—the secretive and oppressive organization that enforced the Catholic faith—to investigate Galileo's assertions and judge whether the Florentine astronomer was guilty of heresy. It was not an audit to be taken lightly; only sixteen years previously Giordano Bruno was burned at the stake for speculating that there might be life on other planets. But Galileo was lucky. The Inquisition found that, though his ideas in *Sidereus Nuncius* were opposed to Holy Scripture, they were not heresy.

Galileo did not stop there, however. In 1632 he published *Dialogue on the Two Great World Systems* wherein he poked fun at the

Pope himself and the geocentric beliefs of the church. Had he been able to resist putting the Pope's own arguments into the mouth of a character named "Simplicio," the history of planetary astronomy might have been very different. Apparently, he could not resist. Any restraint the Inquisition had shown in Galileo's first trial evaporated. The tribunal demanded that he renege his geocentric beliefs and condemned him to house arrest for the rest of his life. Perhaps even more damaging, after *Dialogues*'s publication the Church turned its back on astronomy for well over a century, passing the mantle of planetary discovery to the Protestant countries of Europe.

Meanwhile Jupiter waited, secure behind the high-wire fence of the asteroid belt, for the day when the Starman of Padua's descendants would come looking for it.

## PINBALL WIZARD

It's a funny thing about graduate students: some of them seem to spend most of their time doing nothing but are in fact changing the world. In the early 1960s, a UCLA graduate student named Mike Minovitch worked summers at the famous division of Caltech known as the Jet Propulsion Laboratory. At least, his professors thought he was working there. Minovitch was seldom seen during working hours. The only evidence of his presence on campus was the ever-growing pile of computer printouts that spread like an incoming tidal wave of paper across the floor of his office each morning.

Minovitch was not spending his nights playing computer games. At least, not anything I would recognize as a computer game, although what Minovitch thought is anybody's guess. He was engaged in calculating the trajectories needed to win the solar system's biggest pinball game. In the early 1960s, a scientist named Jim Long and his colleagues in JPL's Future Missions Department had realized that, in fifteen years, Jupiter, Saturn, Uranus, and Neptune were going to align themselves in such a way that a *single* spacecraft had a chance of visiting them all. Laws of celestial mechanics being what they are, this arrangement would begin to drift apart by the end of the 1970s and not return for another 176 years. It was the opportunity not just of a generation but of nearly six generations. The question was, how could a spacecraft achieve the requisite velocity to reach all four planets while they were in the right positions for a rendezvous? The obvious approach—a whopping huge rocket—would not work; it was simply not possible to build a rocket that could accelerate a probe to the velocities needed to reach the solar system's outer regions in time to catch the rare alignment of planets.

This is where Mike Minovitch came in. In 1962 he proposed a practical solution to the problem: getting extra thrust by using a planet's gravity field to produce a slingshot effect. Multiple planets could provide multiple boosts, and quite suddenly the edge of the solar system, perhaps even Pluto itself, became accessible. The answer would be used with notable success just a few years later on *Mariner 10*'s epic journey to Mercury. Just as *Mariner 10* would use the gravity of Venus to slingshot itself to Mercury, Minovitch thought the massive gravity-well of Jupiter would be ideal for accelerating a spacecraft to Saturn, Uranus, and Neptune, picking up additional energy at each of these stops. By the time the probe left Neptune, it would be traveling fast enough to leave our solar system forever. Earth's first interstellar spacecraft was on the drawing board.

In 1965 another JPL alumnus, Gary Flandro, combined the two ideas—the 1970s alignment of the outer gas giants and Minovitch's pinball wizardry—into a plan that became known as the "grand tour" of the outer planets. Using the gravity-wells of Jupiter and the planets beyond it to reach escape velocity, a spacecraft could perform some serious science: it could visit, probe, and image the whole outer solar system. Minovitch, the ultimate pinball wizard, had successfully plotted a course from Earth to Neptune via Jupiter, Saturn, and Uranus. All that was needed to turn a graduate student's nighttime obsession into a reality was a multibillion-dollar spacecraft—or, as it turned out, two spacecraft.

For JPL boss William Pickering, the headaches of the "grand tour" only started with the trajectory. Other problems became all too apparent all too quickly. The largest concerned onboard energy—how to power the electrical systems of a craft heading at an ever-increasing clip away from the sun. The power supply problem was relatively easy with spacecraft investigating the inferior planets Venus and Mercury, and not much more complicated for spacecraft sent to Mars; the gigantic photon stove that is the sun could easily supply enough energy to keep such spacecraft happy via photovoltaic panels. This was a well-tested technology almost as old as the space race itself. But a trip to the outer planets was a completely different proposition. JPL's engineers came up with only two power supply options. The first, supplying the spacecraft with sufficient batteries to keep them running for a decade, was clearly a nonstarter; the rockets would never leave the ground given the necessary weight. The second option was much more elegant, though hardly likely to be popular with the green lobby on Earth. This idea was to supply the spacecraft with what were euphemistically termed radioisotope thermoelectric generators. In other words, the probes would carry their own nuclear power plants.

The eyes of the spacecraft would have to be new, too. Beyond the orbit of Jupiter, the sun would be a bright but distant speck. There would be no natural solar floodlight to illuminate the planets and moons being imaged. Photographs would have to be taken with the camera aperture set wide open and kept that way for minutes at a time. This is easy enough to do when camera and subject are stationary but somewhat harder when both are moving at thousands of kilometers per hour in different directions. A camera system would have to be devised that could compensate for the motion of the spacecraft.

Once the spacecraft's eyes and other instruments were in place, there was the problem of getting its dispatches back to the head office. The radio transmitters JPL had placed on its spacecraft to the inner planets had been simple affairs, with a data transmission rate of a mere eight bps—about what the *Titanic*'s radio operator achieved when sending out distress messages via his Morse code key. For a spacecraft beyond the orbit of Jupiter, trying to take pictures of planetary riches that would remain in range only fleetingly, such a paltry data stream was clearly out of the question. It would take hours to assemble even a single picture. Upgraded communications were going to be a priority.

These were the main problems Pickering had to face. In 1968, years after Mike Minovitch had finished haunting the computer rooms of Pasadena, Pickering called a meeting to discuss turning this vision into reality. Present that day were Homer Joe Stewart, the visionary head of JPL's Future Missions Department, Robert Parks, leader of the Planetary Projects Department, and Clarence Gates, a trajectory specialist. Pickering's concern was, could the project be pulled off? He asked if the "grand tour" alignment was real and was told that it was: the planets would be in an optimum alignment in 1979. He asked how long they would remain aligned and was told that the spacecraft would have a decade to get through to Neptune. He asked what the problems were and was told that the physics and navigation were sorted out and only hardware problems remained.

Pickering told them to get on it.

One of the biggest differences between the Soviet system of space exploration and the American one was that the Russians never worried about changing administrations and the resulting uncertainty over funding. The regimes did, however, share similarities—for example, both liked to pit competing organizations against each other. In the Soviet Union, different divisions of the labyrinthine and creaking bureaucracy under luminaries such as Sergei Korolyov, Vladimir Chelomei, and Valentin Glusko were designed to fight it out for funding and resources. In the United States, it was routine for contractors like Rockwell to face off against General Dynamics, Boeing, and Lockheed. JPL, despite its status as a major NASA contractor, was still just a contractor, subject to the same budgetary constraints as any other.

By 1971 the plans for the "grand tour" spacecraft were completed. They were a monument to Yankee know-how, ingenious enough to make steely eyed missile men from Huntsville to Houston weep with admiration. Nevertheless, it was clear to anybody who understood American space exploration that the project was simply not going to happen. Caught in the middle of a bloody war in southeast Asia, Congress was in no mood to hand out $750 million checks to a bunch of slide rule–carrying Californian longhairs who liked building spacecraft better than helicopter gunships. The "grand tour" was canned.

This was a pity because the spacecraft design was a gem. The plutonium reactors, seven feet long and roughly cylindrical, were designed to lose only 2 percent of their initial charge each year and would power the spacecraft for fifty years—ten times longer than the mission's proposed duration.

Even more impressive was the onboard computer. Previous JPL spacecraft had used the simple but reliable system of a prime processor backed up by a secondary processor. But this design was not adequate for a mission projected to last ten years. Taking the concept of redundancy to a new extreme, the JPL engineers had equipped the "grand tour" craft with five redundant computers—three prime and two backup. Before any critical maneuver was attempted, the three prime computers would consult each other on their calculations. If all three agreed, they would execute the maneuver, but if one disagreed, the two backup computers would be consulted to find out why. If the "minority report" turned out to be correct, then the two backup computers would have the power to implement the decision;

but if the dissenting computer was found guilty of a malfunction, one of the backup computers would take its place. As a computer system it was well ahead of its time, and its designers were justifiably proud of it. They called it STAR—an acronym for Self Test And Repair—but with a name like that, I like to think that the engineers who named it had larger goals in mind than just the "grand tour."

Perhaps the most important part of the system was its ability to communicate with Earth. Everybody was aware that the "grand tour," by far the most ambitious space probe mission ever devised, would be entirely wasted if the images and scientific data it produced could not be returned to the head office in good order. The spacecraft's antenna was a whopping twelve-foot-diameter parabolic affair that would fire data back to Earth not in a trickle, as previous craft had done, but in a torrent of bits. With a data rate of over 115,000 bits per second, the "grand tour" spacecraft could actually talk faster than it could listen, sending real-time data on almost any planetary parameter, from a planet's magnetic fields to the composition of the atmosphere as well as, of course, detailed photographs. The spacecraft would be the first to test the true capabilities of NASA's Deep Space Tracking Network, a network of receiving stations stretching from Australia to Spain to California.

By the early 1970s, the JPL engineers had refined Mike Minovitch's calculations well enough to determine the precise launch window. It would open in mid-August 1977 and close a month later. Minovitch had given them several thousand trajectories to choose from. Indeed, so rich was the choice that, if they wanted, they could even send the spacecraft to have a look at Pluto.

The mission was not shelved for long. Pickering and his team realized their design was too ambitious but saw no reason why something more prosaic would not suffice. They went back to the spacecraft that had made their reputations—the Mariners—and based the new "grand tour" vehicle on that design. Some of the newer innovations remained sacrosanct: the nuclear power source, large antenna, and high-data return rate could not be compromised. But as the 1977 launch window appeared on the horizon, the JPL engineers found that they could accomplish much of their mission with a cheaper pair of spacecraft.

One problem that troubled the hardheaded engineers at JPL was finding a name for their new spacecraft. They thought of *Nomad* and *Pilgrim*, but both names seemed inappropriate to a Congress that was nothing if not literal. Another name made the engineers nervous, for it had already been given to a mission that never made it off the drawing boards—*Voyager*. The original *Voyager*, predecessor to the *Viking* spacecraft, had been canned because, among other things, its design called for two spacecraft to be launched to Mars on the same rocket, an idea that contravened the most fundamental of NASA protocols—multiple redundancy. Putting two spacecraft on one rocket was an easy way to lose two multimillion-dollar budget items with just one mishap. Eventually, though, with Congress insisting on a snappy name for the stripped-down program, superstition gave way to efficiency; the name Voyager was purloined for the mission, and two of the most successful spacecraft NASA would ever launch were readied to shatter the surly bonds of Earth. The mission became a go in May 1972 at about the same time that the scientists of JPL discovered that they had taken their eye off the prize and allowed a competitor to beat them to it.

## THE RADIATION MACHINES

In northern California, the aviators of Moffet Field, home of NASA's Ames research facility, were ready to hurl their own two spacecraft into the depthless tracts beyond the asteroid belt. Ames had always been a competitor in the space program. Incorporated into NASA with the rest of the NACA facilities in 1958, it was run by an individual, Smith DeFrance, whose credentials firmly predated jet aircraft. A former NACA pilot, DeFrance's flying career had ended after he lost an eye during a flight test and promised his wife that he would never climb aboard an airplane again.

Ames was the last NASA facility that anyone would think of as a hotbed of interplanetary innovation, yet it was here that, to a very large extent, the space game was kept alive through the budget cuts of the early 1970s. Smith DeFrance was a conservative administrator, but he knew a good thing when he saw one. As early as 1962 he made the four-day round trip to Washington by train to persuade his bosses at NASA that Ames could build one of the first generation of unmanned interplanetary probes—the *Pioneer*s. The brash young man in charge of the program was Charlie Hall, and when DeFrance was asked if Hall could handle the mission, his answer was unequivocal: "Ames is a hundred percent behind it." It was typical of the can-do attitude that pervaded Ames throughout the space race days of the 1960s and 1970s.

Unknown to DeFrance and his colleagues, one of the reasons Ames received so much early support was NASA's concerns over the prima donnas at the other end of the state. In the early 1960s, JPL was busy with the Ranger series of moon probes, which were to image the moon, transmit pictures back to Pasadena, and then crash into the moon. But the Ranger series was far from totally successful, and NASA was greatly concerned that such a high-profile and expensive project was returning so little data. It was true that JPL, under Pickering's vocal and bullish direction, had put some of America's earliest satellites into orbit, but after that success, JPL had begun to exhibit a worrying excess of personality. To NASA, of course, they were only another contractor. So when Charlie Hall walked into the office of the deputy director of space sciences, Edgar Cortwright, one day in 1962 to talk him into bankrolling the Pioneer program, Cortwright was only too willing to listen to anyone who could reduce NASA's dependence on JPL.

The result of Charlie Hall's initiative was the development of the *Pioneer* probes. All of these craft were of relatively simple construction, reflecting Hall's belief that simplicity is preferable to complexity when designing machinery for deep space. This philosophy, which contrasted sharply with JPL's, explains why it was ultimately Ames that launched the first two probes to Jupiter, *Pioneer 10* and *11*.

In mid-February of 1972, *Pioneer 10* was moved to the gantry at launch complex 36-A at Cape Canaveral Air Force Station. Just prior to the launch, the craft's own twin radioisotope thermoelectric generators (RTGs) were bolted to twin booms that would unfurl when the craft was safely in deep space. These RTGs were as terrifying as they were ingenious. Like the *Voyagers* that would come later, they were designed to keep the spacecraft fully powered for at least fifty years and could only safely be viewed through thickly leaded windows in their isolation facility. Even with their thick casings, the plutonium–238 fuel pellets inside made them too hot to touch. At least one engineer on the *Pioneer 10* team declined to be present at the RTGs' installation on the grounds that he still wanted to have a family.

It was one of the many ironies that surrounded these later *Pioneer* spacecraft that, despite the highly radioactive power sources NASA was about to blast into the skies, most Americans were far more concerned about the line drawings each spacecraft would carry. On both *Pioneers 10* and *11*, Carl Sagan had arranged that a gold-plated aluminum tablet be added as an interstellar greeting card in case either craft should ever encounter an alien civilization. Sagan had been inspired to propose this idea to NASA by Eric Burgess, a writer for the *Christian Science Monitor*. Burgess was worried that, with all the focus on the scientific aspects of the *Pioneer 10* and *11* missions, people might lose touch with the essential wonder of sending a spacecraft beyond the asteroid belt for the first time. It was an excellent point, and it paved the way for the famous golden records that were later carried aboard the *Voyager* spacecraft.

Each of the *Pioneer* plaques, as with the records on the *Voyagers*, contained a visual reference to the Earth's position in space—relative to the fourteen local pulsars—plus a depiction of the timing of the hyperfine transition of the hydrogen atom to introduce aliens to the way we measure time. But far more problematic to the delicate sensibilities of some Americans in the early 1970s were the images of two naked humans, male and female. They had been drawn by Sagan's wife at the time, Linda Salzman Sagan, and approved by NASA administrative mogul John Naugle.

Certain parts of America were outraged, and when images of the plaques were portrayed in the press, the usual approach was to airbrush out the couple's genitals or cover them safely with duct tape. Andrew Khinoy of the *Philadelphia Inquirer* put it succinctly, "What they're ready to accept in outer space, they're not ready to accept in Philadelphia, at least, not on the front pages of their newspapers."

At 8:49 PM Eastern Standard Time on Thursday, March 2, 1972, the mighty engines of *Pioneer 10*'s Atlas first-stage rocket lit up the night sky as they flung the tiny spacecraft toward Jupiter. Less than eighteen minutes later, *Pioneer 10* was moving faster than anything humanity had ever made before, a staggering 32,000 mph. Eleven hours after that, *Pioneer 10* crossed the orbit of the moon and was speeding toward the asteroid belt—the first of many big unknowns the spacecraft would encounter. No human device had ever been through the asteroid belt, and nobody knew whether the rubble would smash the craft or damage its onboard instruments. On February 15, 1973, they had their answer: Ames officially declared that *Pioneer 10* had cleared the belt with no mishaps. It was now on its way to an encounter with the far more hostile environment of Jupiter.

The problem with that environment was radiation, which at least one scientist on the Pioneer team knew that Jupiter supplied by the bucket. James Van Allen had discovered the radiation belt around Earth that bears his name today. That discovery was a direct result of the flight of America's very first satellite, *Explorer 1*, launched during the International Geophysical Year of 1958. Van Allen also discovered that the agitated radio signals being emitted from Jupiter, known since the invention of radio telescopes, were evidence that the region around Jupiter emitted radiation in doses that would kill a person in seconds. With *Pioneer 10* approaching at speeds beyond imagination, the mission planners could only cross their fingers and hope the instruments survived.

Periapsis—*Pioneer 10*'s closest approach to Jupiter—was predicted for Monday, December 3, 1973, at 6:24 PM Pacific time. By the time the spacecraft had cleared the asteroid belt, Jupiter's massive gravity had accelerated *Pioneer 10* to 82,000 mph, and the radiation had risen to levels ten times those predicted. Everyone knew that the imaging photopolarimeter—the spacecraft's "camera"—would go first, and then the other instruments would gradually shut down like the organs of a dying patient. By the time of *Pioneer 10*'s closest approach to Jupiter, everyone believed that the spacecraft would be effectively dead.

But *Pioneer 10* was lucky. Just minutes before the predicted shutdown, radiation levels started to trail off. Unlike Earth's magnetic field, which extends fairly evenly from pole to pole, Jupiter's is a toroidal ring that wobbles around the planet like a spinning plate on a stick. Once the ring moves out of your way, you are back in low-radiation space. For *Pioneer 10*, this happened just in time to stave off immolation. Yet *Pioneer 10* did not escape unscathed. Some of the preliminary images of Io were lost because of radiation-induced glitches, and another instrument failed to restart at all. Still, the craft emerged largely intact. At 6:30 PM on December 3, *Pioneer 10* flew just 81,000 miles above the swirling cloud-tops of Jupiter's roiling atmosphere—and was still taking pictures. The cameras kept rolling, and humanity's first encounter with Jupiter was preserved for posterity. One of the most haunting images in the lexicon of space exploration is the view that *Pioneer 10* sent back as she passed Jupiter and headed into the deepest unknown.

As Charlie Hall pointed out, we were only twelve generations from the person who had first discovered the moons of Jupiter, yet here we were actually visiting the planet. The Starman of Padua would have been proud.

## ENTER THE DRAGON

By today's standards, the pictures returned by *Pioneer 10* are not much to look at. They are grainy and composed of only two colors, red and blue. The additional colors in the press-release photos were added manually. But Tom Gehrels, the inventor of the spacecraft's photopolarimeter, was the first to admit that *Pioneer 10* was not carrying an actual camera. The image photopolarimeter worked by assembling an image strip by strip, a consequence of one of the *Pioneer* probe's principal limitations—spin control stability. Instead of using expensive thrusters, the spacecraft stabilized itself by spinning on its axis. This motion would have made a conventional camera useless. In future missions, as the public's enormous appetite for space images became apparent to NASA, cameras would become overwhelmingly important. Space exploration would then entail striking a delicate balance between science and the public's desire for spectacular photographs of exotic objects. NASA and the other space agencies would quickly become adept at it. And all but the most snobbish researchers understood that the keen public interest was healthy for their careers as well as for planetary science. After *Pioneer 10*, a camera would be a basic requirement for almost all space missions.

The success of *Pioneer 10* was an enormous shot in the arm for the team at Ames and an equally savage kick in the teeth for JPL. Caltech's much vaunted Jet Propulsion Laboratory was no longer the only game in town. An immediate consequence was that JPL tried to take the Pioneer program away from Ames, but they failed.

So successful was the Jupiter flyby that the normally conservative aviators at Ames found themselves thinking, like their Pasadena colleagues, far outside the box. If *Pioneer 10* had fared so well, why not take a chance with *Pioneer 11*? They calculated that a minor tweak of *Pioneer 11*'s orbital trajectory would not only take it past Jupiter but place it on an intercept course for the next gas station down the road—a gas station whose ring-shaped logo put all others to shame: Saturn.

This was not pure vanity. There were sound scientific reasons for going to Saturn. *Pioneer 10* had almost succumbed to the deadly radiation in Jupiter's giant magnetic field, and Ames wanted *Pioneer 11* to get closer still. Rather than risk exposing the fragile craft to levels of radiation that would almost certainly fry its circuits, they decided to modify the flyby so that it traveled almost vertically from pole to pole. The passage through Jupiter's radiation field would be briefer—though more violent—and the path would also make the Saturn trajectory possible. It was ambitious, but the opportunity was too enticing to miss.

The idea that Ames might reach Saturn before JPL was almost too much for Pickering's group to bear. Yet even they had to admit that the idea had a certain logic. Despite the fact that JPL had scaled their mission back after their initial ambitious plans were rejected, their hyper-modified Mariners were still Porsches compared to Ames's Pioneer Volkswagens.

*Pioneer 11* slipped into Jovian space in early November 1974. Unlike her predecessor, she would be flying past Jupiter in a retrograde orbit—against the planet's rotation—which would allow more detailed imaging with the IPP as well as better measurements of Jupiter's magnetic field, taken with a revamped instrument designed by James Van Allen with his characteristic ingenuity. The different orbit would also allow *Pioneer 11* to examine Jupiter's polar regions, where the mission scientists hoped the atmosphere might be more transparent than near the equator.

Instrumentation and approach were not the only things the Ames team did differently. Knowing from their *Pioneer 10* experience that radiation would interfere with the operation of the instruments, they started uploading a continuous stream of compensating commands via the giant antennae of JPL's Deep Space Tracking Network. The tactic worked like a charm, and *Pioneer 11* did not hemorrhage data the way *Pioneer 10* had.

At 9:00 PM Pacific time on December 2, 1974, *Pioneer 11* arrived at its point of closest approach with Jupiter, only about 26,000 miles above the cloud-tops. For the team at Ames, it was an anxious couple of hours, because this time the periapsis would occur above Jupiter's far side and radio communication with the craft would temporarily cease. The Canberra station was the first to reacquire the signal. The period of instrument malfunction had been mercifully brief. The second flyby of Jupiter had been a success, too. NASA official Robert Kramer told the assembled press corps, "*Pioneer 11* flew into the fiery jaws of the dragon and got scorched a little, but it is a tough little bird and is now headed for Saturn."

Before the flights of *Pioneers 10* and *11*, Jupiter had been a largely closed book. Little more was known than that it had moons and a very active radiation environment. The *Pioneer* probes sent home the message that Jupiter is much more complicated than anyone had previously imagined. For example, it is practically all liquid. Its chemistry is dominated by compressed hydrogen and helium, and as far as anyone can see, the planet has no solid core. If there is one, it is likely to consist of something outré—some have even suggested metallic hydrogen, a substance not found on Earth because of the colossal pressures required for its synthesis.

The *Pioneer* probes found that Jupiter's atmosphere is a witch's cocktail of ammonia, methane, and water vapor stirred into a continuous maelstrom that screams eternally across the face of the planet. One of the few features of the planet observable from Earth—the giant red spot that glares malevolently from its equatorial region—turned out to be a purely atmospheric phenomenon, a hydrogen and ammonia hurricane bigger in diameter than the Earth that has been raging for centuries.

The *Pioneer* spacecraft discovered that Jupiter is a net emitter of heat. This internal heat, together with the abundance of hydrogen and other elements on Jupiter, may be responsible for the planet's metallic core, which in turn creates the immense toroidal magnetic field that traps radiation in the planet's vicinity. The Pioneer missions also showed that the Jovian moons play a part in maintaining the complexity of the magnetic and radiation environment around the planet. The larger moons in particular absorb a great deal of radiation and leave trails of reduced radiation in their orbital wakes.

The Ames gamble had paid off. *Pioneer 11* had not only survived its Jupiter encounter but was now headed across the gulf of space to Saturn. The journey would take fully six years, most of it spent above the plane of the ecliptic. The price for the first close look at Saturn was a slow path and a long wait.

## V'GER

In 1979, in one of the most famous franchise extensions in movie history, Captain Kirk, Mr. Spock, Scotty, Lieutenant Uhura, and the others squeeze themselves back into their Lycra uniforms and reboard the *Enterprise* in an effort to save Earth yet again. This time, though, things are serious. Our planet is menaced by a giant spacecraft—indeed, it is more like a small star-cluster—that is inexorably approaching Earth. Duly firing up its refurbished warp engines, the *Enterprise* sprints across the Virgo Supercluster to defuse yet another hostile situation. What no one expects is that the intelligence guiding this ambulatory mini-galaxy has not come from the other side of the universe but from a launch pad in Florida. While en route to the stars, *Voyager 6*—now centuries old—has acquired sentience and decided to come back to Mama. Perhaps taking this movie as a warning, NASA has never dared to launch a *Voyager 6*. In fact, it hasn't yet even attempted a *3*, *4*, or *5*.

*Voyager 1* and *Voyager 2*, however, left Cape Canaveral Air Force Station in the summer of 1977. *Voyager 2* launched first, on August 20, 1977, and *Voyager 1* was launched on September 5. The peculiar reversal of launch order reflected the different mission objectives JPL finally elected for the two craft. *Voyager 1* would concentrate on the two inner planets on the "grand tour"—Jupiter and Saturn—whereas *Voyager 2* was a wild card scheduled to visit Jupiter, Saturn, and, if possible, Uranus and Neptune.

*Voyager 1*'s trajectory had been carefully calculated to allow it to explore one of the most mysterious moons in the solar system—Saturn's Titan—in a flyby so close that it would immediately be flipped out of the solar system, never to return. After a voyage of almost a billion miles, *Voyager 1* would skim Titan's cloud-tops at a distance of 4,000 miles—or from about the same distance as where I write this in my study in West Oxfordshire to Launch Complex 41 at Cape Canaveral Air Force Station in Florida from which the *Voyager*s departed.

*Voyager 1*'s path was made shorter and, therefore, faster than its sister's to enable it to arrive in the outer solar system on time to meet Titan. *Voyager 2*'s more leisurely path would allow it to fly on to Uranus and Neptune or, if necessary, to be redirected if *Voyager 1* failed in its Titan encounter.

Titan was considered important because Earth-based spectroscopic examination showed that its atmosphere was composed largely of primitive organic molecules. This finding stimulated Carl Sagan to promote Titan as a place where life might exist, putting it squarely on the map as a "must-visit" destination for at least one of the *Voyager* probes. It is a measure of Sagan's influence on American space science that JPL was willing to redirect its second *Voyager* probe if the first one failed to reach Titan and forego the chance to visit Jupiter and Neptune.

The "grand tour" orbital alignment meant that it would take only two years for *Voyager 1* to reach Jupiter and a little more than three for it to reach Saturn. Compare that to the six-year odyssey that *Pioneer 11* was enduring—and that was six years from Jupiter. It was one of the ironies of solar system exploration that, by the time *Pioneer 11* reached Saturn, *Voyager 1* would be hard on its heels.

The decision to send *Voyager 2* to Uranus and Neptune was largely a result of the unexpected longevity of *Pioneer 11*. The JPL engineers found themselves wondering what would happen if either or both of their *Voyager* craft exhibited such happy tendencies. It was one thing for Washington to refuse to spend money on a spacecraft *designed* to reach Uranus and Neptune, but it was quite another simply to let the craft go there. Once launched, missions were measured in years—sometimes in decades—and quite often lasted longer than the duration of any one political administration. It seemed quite possible to the engineers at JPL that, by the time *Voyager 2* reached Jupiter, the political winds might have shifted. It was not impossible that funding for a mission extension would be forthcoming—providing the spacecraft was up to it.

With all this in mind, the Voyager engineers had designed their socks off. Nothing illustrates the care with which JPL designed its birds than the fact that after departure the engineers continued working on the software and, via remote-control reprogramming, enhanced the capabilities of both spacecraft en route. With such flexibility built into them, the *Voyager* spacecraft were set to revolutionize planetary astronomy. *Voyager 1* arrived at Jupiter in the spring of 1979, and *Voyager 2* followed in the summer. *Voyager 1* made its closest approach to Jupiter on March 5, 1979, at 128,400 miles, and *Voyager 2*, on July 9, 1979, at 350,000 miles.

An early priority for both *Voyager* craft was to image Io, since *Pioneer 11* had noted a red tint to the moon on its way through the Jovian system. The early photographs sent back to Earth produced a sensation. Io has active volcanism—the first world in the solar system apart from Venus and our own to exhibit this trait. The discovery, made by navigation engineer Linda Hyder at JPL, sent shockwaves through the planetary science community and was the first hint of the incredible diversity of Jupiter's moons. Further imaging showed that Io is pockmarked with volcanoes and yet has almost no craters. The clear implication is that Io, much like Venus, is being continually resurfaced by volcanism.

The *Voyager*s observed a total of nine volcanic eruptions on Io, and it is clear that other eruptions occurred between the two spacecrafts' encounters. But volcanism on Io is not the passive affair it is on Earth, where the planet's atmosphere quickly dissipates the ejecta. Plumes from Io's volcanoes tower more than two hundred miles above the surface, escaping the moon's weak gravity and coating the Jovian planetary system in a dust of sulfur, oxygen, and sodium. Io's volcanism is apparently caused by a phenomenon at the heart of the geology of the Jovian moons—tidal kneading. Jupiter is so massive and other moons such as Europa and Ganymede pass by so closely that Io is continually stretched and distorted as it orbits Jupiter. The three moons are locked into what is known as a simple Laplace resonance of 1:2:4, where the orbit of each is largely dictated by the orbit of the others. One of the symptoms of this cosmic weight-lifting is that Io's tidal bulge is a staggering one hundred times greater than Earth's; it is 330 feet compared to just three. Io is now considered the most volcanically active moon in the solar system. Indications of its extreme temperatures are found in the gouts of molten silica that spew from its ubiquitous volcanoes.

Although Io was interesting, it was Europa that caused the most head scratching. *Voyager 1* photographed it as a blue-white ball crisscrossed with black lines. The relatively low-resolution pictures sent back by *Voyager 1* did not show clearly what these lines might be. The initial thought at JPL was that they were caused by crustal fracturing or related tectonic processes. But as *Voyager 2* came in for a closer look, the mystery deepened. The features had so little relief that they looked as though they might have been applied with a paintbrush. It was the start of a major change in our understanding of the nature of moons and planets of the solar system, because Europa's surface turned out to be another manifestation of the Laplace resonance and associated tidal kneading that torments Io. Europa is just far enough away from Jupiter that she is not gravitationally pummeled like Io, and just close enough that she is not inactive like Ganymede and Callisto. Her blue-white color turned out to be due to a sheath of ice—many kilometers thick—that completely covers the surface. The lines are the cracks between blocks. The *Voyager* instruments showed that Europa has a differentiated interior with a two-layer core, a mantle of relatively warm ice, and an outer crust of brittle ice. Beneath that ice, the Voyager scientists suspected, is a layer of water—ordinary liquid water—above the moon's silicate core. Since the Jovian system was known to be awash with the same type of elements and compounds that inhabit the inner solar system, the possibility that life exists in the oceans of Europa could not be discounted. But investigating that issue would have to wait for another, appropriately equipped probe.

The *Voyager* imaging of Ganymede showed that it is the largest moon in the solar system, with a diameter of over three thousand miles. Unlike Io and Europa, Ganymede's surface is divided into two distinct types of terrain, cratered and grooved, which suggests that its icy crust is also continually pulled by tidal kneading. If the moon were unleashed from Jupiter's massive gravity, Ganymede would be a planet in its own right, and a complex one at that. It is one of the bodies of our solar system that blurs the definition between moon and planet.

Callisto is more like what was expected. Its heavily cratered surface implies that it is very old. Yet it, too, shows the influence of ice. The very largest craters are apparently worn down like ancient teeth by the passage of a million glaciers. Only ghosts of these craters are left, identifiable by a lighter hue and the faint traces of surrounding concentric ridges. One of the most extraordinary structures on Callisto is a line of craters that look as though the surface of the ancient moon was strafed with machine-gun fire. The specialists at JPL hypothesize that Callisto was once hit by an asteroid that bounced across its surface like a pebble skipping across a pond.

The *Voyager* spacecraft found that Jupiter has a faint system of rings, although only a pale shadow of those orbiting her next-door neighbor, and discovered three new moons: Adrastea and Metis orbiting just outside the rings, and Thebe orbiting between Amalthea and Io. The *Voyager*s also confirmed the intensity of the Jovian radiation environment, which, fortunately, JPL had learned from the *Pioneer* example and thus sent off its craft well armored. Together with Io's intense volcanism, this radiation storm is partially responsible for stripping material from the surface of that moon. Io also experiences intense electrical activity as it moves through the Jovian magnetic field, developing 400,000 volts across its diameter and generating a 3 million-ampere electrical current that flows down the flux lines of Jupiter's magnetic field and into the planet's ionosphere.

In the late summer of 1979, the two *Voyager* spacecraft pulled away from Jupiter and started the long, lonely trek to Saturn. Meanwhile, enough had been learned back on Earth to justify another mission to the eye of the universe.

## THE STARS MY DESTINATION

There is no denying that the Voyager missions rank among the most successful in the history of space exploration. But before the two vehicles even left the ground, COMPLEX (the successor to the fabled Committee 14 of the National Academy of Sciences, which provides advice to NASA on space exploration strategy) had already recommended that the Voyager reconnaissance missions be followed by a detailed investigation of Jupiter with a spacecraft fine-tuned for the

Jovian environment. As the *Voyagers* were pulling away from the Jovian system and heading toward Saturn, the new spacecraft—to be named *Galileo*—was under construction. *Galileo* was initially scheduled for launch from a space shuttle in 1982, with arrival at Jupiter in 1985. The shuttle would carry a Centaur second-stage booster to low Earth orbit, enabling the new spacecraft to sprint across the solar system and arrive at Jupiter only six years after its two *Voyager* cousins had departed.

At the time this decision was made, the space shuttle was still new. Its first successful flight had occurred in 1981—only two years after the *Voyager* spacecraft arrived in Jupiter orbit—and the choice of the space shuttle/Centaur combination for the *Galileo* launch reflected not only a breezy optimism with America's new toy but also a determination to justify the billions that had been spent on its development.

I can still remember the first space shuttle launch. It was April 12, 1981, and I was home from college visiting my folks. The shuttle was *Columbia*, the spacecraft that would break up over Texas on re-entry in 2003. But this was 1981, and the future was still bright and full of hope. As I watched the television with my parents, that distant white bird on the pad at Launch Complex 39A suddenly shook as geysers of smoke and flame exploded through the water-drenched, vibration-quenching mechanisms of the pad. The noise built, and then quite suddenly she was off, under the sure control of test pilot astronauts John Young and Bob Crippen, hurtling upward into the twilight of the thin upper atmosphere.

A hugely ambitious spacecraft, *Galileo* comprised two sections: an orbiter and a probe. At launch the two together weighed over 2,500 kilograms and stood seven meters tall. The spacecraft's lineage from its *Pioneer* and *Voyager* ancestors was obvious in that it possessed both a rotating section (for stability and the benefit of scanning instruments such as magnetometers) and a de-spun section for the cameras and other instruments that benefited from stability. *Galileo* was a collaborative effort, built by the missile scientists of JPL, with Germany supplying the propulsion module and Ames—in recognition of its pioneering achievements—tasked with the responsibility of managing the probe. The craft's prime mission was to study the Jovian system for two years. She would swoop around the eye of the universe in long, two-month ellipses, getting the measure of Jupiter's colossal magnetosphere as well as performing several close flybys of Jupiter's largest moons to expand on the tantalizing hints sent back by the *Voyagers*.

The construction of such an ambitious spacecraft was fraught with delays, and *Galileo*'s launch windows slipped past, one by one, as shuttle after shuttle hurled itself successfully into Earth orbit. Years passed; 1982 slipped to 1985 and then to 1986, and then the nightmare that had haunted NASA for almost thirty years struck. At 11:39 AM Eastern Standard Time on January 28, 1986, *Challenger* exploded seventy-three seconds after launch, killing all seven of its crew.

The *Challenger* disaster froze the shuttle program and paralyzed America's confidence in space exploration. Not until 1989 were shuttle flights resumed under much more stringent safety conditions. The condition that most immediately affected the Galileo mission was that the mighty hydrogen-fueled Centaur rocket that was to fling the spacecraft across the solar system to Jupiter was deemed too dangerous to haul aloft in a manned vehicle. The lower-powered rocket that had to be used instead would not get *Galileo* to Jupiter until 1995, by which time *Voyager 1* would have flown by Titan on its way out of the solar system and *Voyager 2* would be crossing the trackless gulfs of deep space to Uranus and Neptune.

Controversy surrounded *Galileo*'s launch. Once again, it was determined that the only reasonable form of power for such a deep-space craft would be the RTGs that had performed so successfully on the Pioneer and Voyager missions. But this time, as the date neared, antinuclear protest groups sought a court injunction prohibiting *Galileo*'s launch. The *Challenger* explosion had shown the American public that catastrophic launch failures could happen, and nobody wanted several kilos of plutonium–238 raining down on their heads if this launch vehicle should fail, too. Furthermore, the lower-powered booster that *Galileo* would now be using meant that several gravity assist flybys of Venus and Earth were called for, and few people relished a radioactive spacecraft slingshotting itself around our home world. These were legitimate concerns—especially in the light of NASA's incompetence as displayed by the *Challenger* disaster—and serious people did not treat them with contempt. Carl Sagan himself, one of the strongest and most vocal supporters of the *Galileo* mission, went as far as to say that "there is nothing absurd about either side of this argument." His message was clear: the stars, yes—but not at any cost.

*Galileo* arrived at Jupiter in December 1995, skimming close by Europa and Io on the way. By July the descent probe had completed its five-month fall toward the planet's roiling cloud-tops. Since the probe had no maneuverability of its own, the main orbiter had been aligned carefully to ensure the probe's successful rendezvous with the planet. Just before entry, the orbiter started receiving data from the descent probe, which sliced into the top of the Jovian atmosphere at over 100,000 mph, was slowed by aerodynamic braking, deployed its parachute, and then released its heat shield. For nearly an hour it transmitted data as it swung downward through the maelstrom of Jupiter's atmosphere. Eventually, temperatures equivalent to that of a domestic oven and pressures similar to those found at the bottom of Earth's oceans destroyed the probe, but not before its mass spectrometer had sent back the surprising information that Jupiter's atmospheric composition was radically different from what was expected.

Mission scientists had speculated that Jupiter's atmosphere—like the atmospheres of the other planets—had formed simultaneously with the bulk of the planet as it condensed out of the solar disk. This was not what the probe found, however. Helium, for example, was far less abundant than expected, while there was much more hydrogen. Perhaps the core of Jupiter is composed of metallic helium rather than the marginally lighter hydrogen? Neon, too, was much less abundant than predicted and may also be sequestered in the planet's core. Perhaps most surprisingly of all, carbon—the building block of life—was staggeringly abundant. How could this be? One possible explanation is that Jupiter's massive gravity has, over the eons, attracted a heavy rain of icy comets that are well known to be rich with organic materials. Almost as soon as it had arrived, *Galileo* was rewriting the Jupiter rule book.

The descent probe's findings suggest that Jupiter's atmosphere evolved through a complex and as yet little-understood set of reactions *after* it had condensed out of the solar nebula. The probe also showed some amazing detail of the planet's atmosphere. Jupiter has massive thunderstorms, which are largely restricted to narrow belts above and below the equator. In these belts, wind velocities are extreme and very turbulent, and lightning strokes can be up to one thousand times more powerful than those on Earth.

The first mission phase of the *Galileo* orbiter focused on the moons, and the data it sent back had even more extraordinary implications than finding carbon in the atmosphere of Jupiter itself. These findings were confirmed and amplified by a series of daringly close flybys—particularly of Io and Europa—after the mission was extended because of a phenomenon NASA and its contractors found increasingly familiar—extreme spacecraft longevity.

Magnetic data collected by *Galileo* provided indisputable evidence that not only Europa but also Ganymede and Callisto have liquid saltwater oceans below their frozen surfaces. Indeed, the only one of the original Medicean stars *not* to possess liquid water is volcanic Io. It is an astounding finding, one that even the Starman of Padua himself would probably not have dreamt of. With all that carbon around, the possibility becomes greater that at least one and perhaps several of the moons of Jupiter hold at least the basic building blocks of life.

The *Galileo* orbiter also showed that Europa, Io, and Ganymede all have metallic cores of separated dense metals. Only Callisto appears to be undifferentiated. Europa, Ganymede, and Callisto all have thin atmospheres, known technically as "surface-bound exospheres," which are composed of electrically charged gases loosely bound to the surface. Ganymede, like Earth, has its own magnetic field—the only moon in the solar system known to have one.

The extended mission was harrowing for the small craft. The flybys of highly charged Io eventually damaged some of its critical systems. But it was enough. *Galileo* had already performed far beyond expectations, and the images of the comet Shoemaker-Levy 9 crashing into Jupiter, which it sent back shortly before it was destroyed, are among the most extraordinary deep-space pictures ever taken.

Eventually, partially blind and with many of its instruments malfunctioning from overexposure to Jupiter's malevolent magnetosphere, the *Galileo* spacecraft was de-orbited in 2003 and allowed to crash into Jupiter. But its legacy cannot be overestimated. By verifying the extraordinary diversity of the Jovian system, it has confirmed what many had begun to suspect—that the Jovian system is a miniature version of our solar system, with as many different types of moons as there are types of planets.

Meanwhile, far beyond the orbit of Jupiter, the *Voyager* probes were making more startling discoveries of their own.

# 8

# Ringing Infinity: Saturn

There is a picture on an office wall in a modern building in the modern British city of Milton Keynes. It is one of the most extraordinary pictures I have ever seen. It shows a small spacecraft adrift in a realm of planets and moons. It could be anywhere in a cosmos of infinite possibility except for the planet that dominates the picture, a planet more recognizable than any other in our already remarkable solar system. This planet is surrounded by a series of concentric filigree rings that girdle its equator. It is heartbreakingly beautiful—it is Saturn.

The rather spartan office is that of Professor John Zarnecki—current director of CEPSAR—the Centre for Earth, Planetary, Space and Astronomical Research at the Open University. The picture is there to remind John that he has spent a good proportion of his working life trying to understand Saturn and, in particular, its moon Titan. And, indeed, John has done more than most to enhance our understanding of Titan, because he has landed there. Not in person, you understand, but as good as in person, for John is one (he is too modest to claim more than a tiny portion of the credit) of the architects of one of the most successful planetary missions of the past decade, the successful landing on Titan. In January 2005, a tiny wok-shaped probe built a decade before successfully parachuted through Titan's atmosphere and landed on a world similar to our own Earth 4 billion years ago. John Zarnecki, you could say, built a time machine.

## OTHER DAYS, OTHER EYES

Six months after he discovered the moons of Jupiter, Galileo discovered Saturn's most characteristic feature; it was July of 1610. His sketches show that he noticed the strange structures that cling so closely to the planet's bulk. But Galileo remained perplexed by them, believing they might be stationary moons—moons that did not orbit about their master as the moons of Jupiter did. Strangely, too, for such an assiduous observer, he did not detect the small bright speck that hung further away from the disk of Saturn and that did rotate about it. Galileo had arrived at the limitations of his instrument, and although the contributions he made to planetary science remain incalculable, it would be another thirty years or so before the mysteries of Saturn would begin to be unfurled.

Christian Huygens was born in 1629 in The Hague in Holland. As a young man, he studied law before eventually turning to science. Huygens was a child of his time; a scientific revolution had begun in 1543 when Copernicus published his theory of orbiting bodies, and Andreas Vesalius had put the mysteries of the human body on a scientific footing by creating the science of anatomy. Like many scholars of that time, Huygens was not afraid to get his hands dirty by making his own instruments, and in collaboration with his brother Constantin, he designed a machine that used gears to make lenses. In a single stroke, the Huygens brothers advanced the science of astronomy by a quantum leap when, by combining precise mathematics and unwavering mechanical cutters, they produced lenses of far greater precision than anything previously made. In conjunction with an ingenious system of ropes and pulleys, the Huygens brothers exploited their superior lenses by constructing telescopes of very long focal length. In 1655, using his strange "aerial" telescope, Christian Huygens discovered what Galileo with his lesser technology had missed—the speck that orbited Saturn, the moon that would be known as Titan. Observations on subsequent nights convinced Huygens that this dim speck was indeed a moon that completed a circuit around Saturn once every sixteen days. Huygens's extraordinary telescope also enabled him to correctly deduce the nature of the strange structures that had so perplexed Galileo. They were not stationary moons; they were rings.

Huygens did not name the moon; he simply referred to it as "Saturn's moon" and soon turned his attention to other astronomical matters. In 1659 Huygens made history when he became the first person to identify a permanent feature on another planet. Syrtis Major, the "Great Marsh," was thought by Huygens and many who came after him to be some kind of watery body on Mars. We should not sneer at Huygens for this assumption. While I was writing this book, NASA announced that *Mars Global Surveyor* had discovered traces of water that had been liquid on Mars within the last five years.

Huygens was a consummate scientist who was intimately involved in the development of calculus and passionately interested in the development of accurate timepieces. He quickly used his discovery of Syrtis Major to prove that the time Mars takes to rotate once on its axis—its day—is similar to the length of our own.

After Huygens turned his attention from Titan, the cudgels were taken up by a young Italian named Giovanni Cassini who, in 1669, was appointed director of the Paris Observatory by King Louis XIV. By 1672 Cassini had discovered two other moons of Saturn, Iapetus and Rhea, and by 1684 he had brought the total of known moons to five with his discovery of Dione and Tethys. Cassini was the first to really study the rings of Saturn and to draw attention to the narrow gap between two of the rings, which even today bears his name—the Cassini division.

The next person to make a significant contribution to the study of Saturn was William Herschel, the musician turned astronomer who discovered Uranus in 1781. In 1789 the rings of Saturn were orientated such that they were edge on when viewed from the Earth. This made for much clearer viewing of the planet, and Herschel was able to detect for the first time two inner satellites that he named Mimas and Enceladus. By making measurements of the positions of these and the other moons, Herschel calculated the duration of their orbit around Saturn. Given these and other measurements, by the 1840s it was determined that Titan orbited Saturn in just under sixteen days.

The first suspicion that Titan might have an atmosphere came in the early twentieth century when the Catalan astronomer Josep Comas i Sola observed Titan through the thirty-eight-cm telescope at the Fabra Observatory near Barcelona, Spain. He noted that Titan had darkened edges, which he attributed to the presence of a dense atmosphere. It was a controversial suggestion at the time, made doubly so by Titan's extreme distance and small size. Support came from a surprising quarter, however, when the noted British astrophysicist Sir James Jeans reasoned that, on chemical grounds, Titan was large enough to retain an atmosphere. Jeans pointed out that, in the frigid depths of the solar system where the kinetic activity of molecules is low, Titan's gravity would be strong enough to have retained an atmosphere since the time of its formation. Jeans's calculations showed that, although light gases such as hydrogen and helium would escape, heavy molecules such as argon, neon, and even nitrogen and methane could still be present. It was an extraordinary suggestion, and one that, as we shall see, had profound implications.

The next major advance in understanding Saturn's largest moon came with the development of the science of spectroscopy. Spectroscopy is a technique for identifying different chemicals remotely by using light wavelength and intensity signatures. In the winter of 1943 to 1944—during the dark days of World War II—the Dutch American astronomer Gerard Kuiper used the newly constructed telescope at the McDonald Observatory in Texas to analyze the spectra of the ten largest moons in the solar system. It was an exercise that had not been attempted before, and Kuiper assiduously collected data in both the visible light and infrared parts of the electromagnetic spectrum. Kuiper immediately noticed that Titan's spectrum was radically different from the rest. It contained characteristic absorption bands for a gas we are readily familiar with—methane. The presence of methane also explained something that had been perplexing planetary astronomers for some time—Titan's baleful orange color. Kuiper was not slow in recognizing the implications of his measurements and published a paper titled "Titan: A Satellite with an Atmosphere." It was proof of Sir James Jeans's theory that, not only did Titan have an atmosphere, but it was an atmosphere with heavy organic molecules in it.

It was now clear that Titan was quite different in many ways to the other satellites of the solar system. It was enough to intrigue the scientific community, most especially the young Carl Sagan. Sagan lobbied extensively for the *Voyager* spacecraft to look closely at Titan. Consequently, *Voyager 1*'s trajectory was specifically aimed at this moon, while *Voyager 2*'s route was held ready for last-minute modification in case *Voyager 1* failed. There can be no clearer indication of the importance that was attached to Titan than that it was so close to the top of the priority list for the Rolls-Royce of all unmanned missions to date.

In the meantime, however, a spacecraft nobody thought would get anywhere close to Saturn was approaching it.

## THE NEEDS OF THE MANY

I was twelve when *Pioneer 11* departed the Jupiter system. During the bulk of my teenage years, *Pioneer 11* had been arcing high above the plane of the ecliptic, heading toward an appointment with destiny in the outer solar system. It—and its sister ship *Pioneer 10*—were now the fastest-moving, most distant human-made objects in the universe.

In the United States, things had changed quite a bit around the Pioneer mission office at Ames. Things there were running down, and the glory days were over. They were working on one final mission—the *Pioneer Venus* probe discussed in Chapter 3—but that would be the last of the Pioneer missions. The people at Ames had come to appreciate something that their Pasadena colleagues would come to know only too well themselves—that space missions are like extended wars with long (very long) periods of boredom punctuated by fleeting moments of sheer terror. With *Pioneer 11* approaching Saturn, however, Charlie Hall's swansong had come. It was to take the first close-up look at the ringed giant.

But *Pioneer 11*'s status had changed in those six years as well. Now NASA was openly referring to both the Pioneer 10 and Pioneer 11 missions as "reconnaissance missions" for its much more expansive Voyager program. This change in emphasis began to affect what *Pioneer 11* would do when it eventually reached Saturn.

For much of the twentieth century, Saturn's ring system was thought to be composed of three concentric rings: the outermost "A" ring, the middle "B" ring, and the faint "C" ring closest to the planet. The Cassini division separated the outer two rings. But less than a decade before the *Voyagers*' launch, a French astronomer named Pierre Guerin claimed to see a fourth ring *inside* the third ring. This ring was made of a material so nebulous that it could barely be discerned by the most powerful telescope of the day, and it was separated from the C ring only by a gap that some wags referred to as the "French division."

The debate about the nature of Saturn's rings was not trivial to the Pioneer and Voyager mission planners; it had practical implications on an engineering level. If the particles in the rings were smaller than about a millimeter, then any incoming craft could pass safely through them. If they were larger than a centimeter, then in all probability the particles would be spaced sufficiently for the spacecraft to get through, once again, unscathed. If the particles, however, were larger than a millimeter but smaller than a centimeter, the spacecraft could find itself in trouble. In 1976, the path of *Pioneer 11*, hitherto an arcane issue of celestial mechanics of interest only to mission specialists, suddenly became a public debate. Any course adjustment would have to be made by the middle of 1978 at the latest. By the beginning of 1977, the debate had condensed into the so-called "inside option" and the "outside option."

In the end, the problem was focused on one thing—the discovery of a fifth ring outside the others, the so-called "E ring," which might extend as much as twice the diameter of the then known total Saturnian ring system. If the outer ice giants were to be targets, then *Voyager 2* would have to pass very close to the edge of the A ring to slingshot itself on to Uranus. In other words, it would have to pass through the E ring. The question for NASA, therefore, boiled down to whether to send *Pioneer 11* through the debris-free gap between the innermost ring and Saturn itself and use *Pioneer*'s aging instruments to image the planet, or whether to use *Pioneer 11* as a guinea pig and send it through the hypothetical E ring to see if it sustained any damage.

The Ames group wanted to send their plucky little *Pioneer 11* out in a blaze of glory by sending it as close to the planet as possible. The "inside track" might provide valuable information about the nature of Saturn—for example, the way the radiation belts and magnetic fields of the rings interacted with charged particles in the Saturnian system. Given the existing *Voyager* flight plan, this information would not be forthcoming due to the more distant flyby.

But the Ames people were also realists. They knew that any observations from beyond Jupiter were gravity gravy. And JPL's steely eyed missile men had a $1 billion argument—if *Pioneer 11* did not survive its passage through the outer rings, then *Voyager 2* could not safely try for Uranus, and it could stay in the Saturnian system doing the work that *Pioneer 11* would have done. It was a difficult argument to counter.

The position of the Pioneer scientists—the principal investigators—was influenced by the experiments they had onboard the craft. Particle and magnetic field specialists such as John Simpson and James Van Allen wanted to go with the inside track option because such a trajectory would provide evidence about whether Saturn had a magnetic field. They believed that *Pioneer 11*'s destruction was worth the possibility of new discoveries.

Yet, this view was not shared by all the Pioneer mission scientists. Dawning on them all was the recognition that deep-space research was a collaborative enterprise. The heady days of the 1960s when space science money was running in the gutters were long gone. They all knew that to make progress they would have to work together. Partly this view recognized the duration of these deep-space voyages—years piled upon years that could easily eat half your career. The only sensible way to make such a sacrifice was to be a team player. The needs of the many outweighed the needs of the few, or the one.

The man in charge of the decision was a NASA official named Thomas Young. His decision was not an easy one by any evaluation, but in the end, Young finally announced his choice: *Pioneer 11* would take the outer track. It would try to blaze a trail through the E ring for *Voyager 2*. Young explained his reasoning in a letter to Charlie Hall. He had, he said, been impressed by the Pioneer scientists' arguments and could appreciate why many favored the inside option. But he had chosen the outside option because the long-range aims of solar system exploration had to outweigh the immediate preferences of an individual project. It was essential to do everything possible to maximize *Voyager 2*'s success. If *Pioneer* failed at the E ring, *Voyager 2* would have to abandon Uranus, and that information was needed now. Once again, the needs of the many outweighed the needs of the few.

And so, during 1979, the year that the *Voyager* spacecraft reached Jupiter, *Pioneer 11* was hurtling through the unknown dangers around Saturn. Tom Gehrel's IPP began taking images of Saturn in early August 1979. The first images were somewhat disappointing, but they improved as the spacecraft got closer. Soon *Pioneer 11* was able to make out more details of the rings, and suddenly Saturn became like Jupiter—a place rather than a dot in the sky. If you had to pin down one legacy for the Pioneer missions, it must surely be this—they made the outer planets real. By early morning of September 1, 1979, the IPP had identified another, previously unknown ring around Saturn. This was labeled, with dry humor, the "F" ring. The IPP also sent back images of a previously undiscovered moon.

At the same time, tension was rising at mission control because later that same morning would come the acid test—when *Pioneer 11* crossed the plane of Saturn's rings. By this time the spacecraft was traveling over 70,000 mph and would pass through the danger zone in less than a second, yet another example of the strange nature of planetary exploration—a six-year journey put at risk by a one-second encounter. There was nothing anyone could do at this point; the round-trip time for a signal to and from *Pioneer 11* was now almost three hours. But at just a few minutes past nine o'clock in the morning, the Ames computer screens began to flicker once again, and the waiting scientists knew that *Pioneer 11* had survived the first ring crossing.

*Pioneer 11* swung past Saturn at a distance of only 13,000 miles, and then communications blacked out as it passed behind the planet.

It would swing back through the rings in the blackout region. But, once again, the craft survived. *Pioneer 11* had survived two ring crossings, and as a direct result, Uranus and the edge of the solar system were now within *Voyager 2*'s reach.

*Pioneer 11* passed closely by Titan on September 2, sending back the first close-up pictures of that moon, then slid quietly out of Saturnian space and into the history books.

## HAND-BRAKE TURN

Can you imagine what it must be like to spend half a billion dollars to launch a spacecraft and then, after 800 million miles, have it turn a corner with a radius of only 4,000 miles? It is like flying from London to New York and then turning around on the point of a pin. It is a piece of high-speed maneuvering that simply had no parallel on Earth.

So it was that, on the evening of November 12, 1980, the *Voyager 1* mission engineers were very far from relaxed. They were only too well aware that a half billion dollars might be flushed down the tubes by the end of the night. In the two years since *Voyager 1* had left Jupiter, the spacecraft had been traveling at roughly 35,000 mph and had put an additional 400 million miles on the odometer. At that speed, the ship sent to do Carl Sagan's bidding at Titan would only have time for one close-up but fleeting look.

In fact, not all the engineers were convinced of the wisdom of the Titan encounter, and *Voyager 1*'s new mission manager, Ray Heacock, was among their number. Heacock had commanded *Voyager 1* and *2* through their Jupiter encounters and had then been responsible for laying in the trajectories that would take both ships to Saturn. After the success of the Jupiter flybys, the Voyager missions had taken on almost mythical status in the hallowed halls of NASA, and Heacock did not want to be the first in Voyager's long line of mission managers to wreck a ship.

The Titan maneuver called for the spacecraft to pass underneath Titan's south pole, then allow itself to be grabbed by the moon's gravity and fired northward up the opposite side. The craft would take readings the entire time before being flung out of the plane of the ecliptic, never to return. As *Voyager 1* approached to within 4,023 miles of Titan's surface, it burned its engine continuously to keep its speed high and its direction true. Years of work and billions of dollars were now quite literally riding on the spacecraft's internal guidance, for instructions from the engineers—even at the speed of light—took over seventy minutes to reach the ship.

Although the maneuver was complicated, there were some things, at least, that were going for it. Since Titan was currently below the ring plane of Saturn, the spacecraft would not have to fight its way across that particular debris field. The southern polar approach meant that collision hazards would be minimized and that any problems that did crop up after the spacecraft's handbrake turn would occur after the machine had fired its priceless data back to Earth.

After Heacock had instructed the flight dynamics officer to issue the necessary commands, there was nothing to do but wait. Sir Isaac Newton was in the driving seat now. If anything went wrong, they would not know about it until long after the planned flyby. A few seconds after the expected time, the screens lit up as a torrent of data cascaded across them. The maneuver had been a success. There was high excitement in their modest meeting room as the imaging scientists waited for the first high-quality, close-up pictures of Titan to come in. Although the pictures would initially be monochrome, these people were as skilled in their interpretation as the best military aerial reconnaissance experts. The color would be added to the images later, after analysts had decoded the chemical spectra of Titan and inferred from them the necessary hues.

As the monitor snow began to dissipate, the faint outline of a ghostly sphere began to coalesce in the flickering phosphorescence of the screen. Gradually it sharpened into a featureless ball. Around the table was absolute stillness as face after face fell. The moment that they had gone to so much trouble to engineer—the moon that Carl Sagan said was one of the most important places in the solar system—was as featureless as a billiard ball. Not only did it have no surface topography, there wasn't even as much as a swirl in its cloud-shrouded surface. Titan was about as exciting as a foggy day in Pittsburgh.

It was one of the worst moments in the history of unmanned space exploration. Another billion bucks shot, and for what? People were well aware that only four years earlier, the two *Viking* spacecraft (another Sagan-backed project one couldn't help noting) had landed on Mars and had singularly failed to find anything as exciting as a microbe.

But such harsh criticism is unfair, for how could we find out that Titan was potentially disappointing unless we went there to see for ourselves? It is a minor flicker in the first flush of space-age optimism. Like the other ambitious unmanned space missions, Voyager was a success in demonstrating that such long voyages were achievable in the first place and by stimulating follow-up missions. In the meantime, only a few months later, *Voyager 2* arrived in Saturn space for its own close encounters.

## CLOSE ENCOUNTERS OF THE TEFLON KIND

Despite the Titan disappointments, the rest of *Voyager 1*'s visit to Saturn was a resounding success. Less than a day after the Titan flyby, the spacecraft passed within 250,000 miles of Tethys. In short order, Mimas, Enceladus, Dione, Rhea, and Hyperion followed. Each encounter sent back valuable information to JPL, and when, in August 1981, *Voyager 2* finally reached the Saturnian system, the JPL engineers used the information to reprogram the ship to take the best advantage of her sister's reconnaissance. Having seen Titan close up and personal, JPL was in no mood to take another swing at the ugly-duckling of the solar system and allowed *Voyager 2* to slip disdainfully by at a distance of almost half a million miles.

*Voyager 1*'s observations suggested that Enceladus and Tethys were interesting subjects for a closer look. Their proximity to each other implied that they might be subjected to tidal kneading in the same way that Europa, Ganymede, and Io are in the Jovian system. Enceladus's surface was so bright that it seemed likely that the moon was active—perhaps even being volcanically resurfaced in a similar fashion to Venus at the other end of the solar system.

Tethys turned out to be significant for just the opposite reason. It has a tormented and apparently very ancient terrain that suggests that it has *no* internal tidal kneading of its own. Indeed, Tethys, which measures less than 700 miles across, is scarred by a massive crater that stretches nearly three-quarters of the way around its circumference. The size of the scar indicates that, at some point in its history, Tethys was hit by one heck of a projectile. Given a bigger hit and the moon would have shattered like a Ming vase under a jackhammer.

After a journey of almost 1 billion miles and four years, *Voyager 2* would be within the Saturnian system for only twenty-one hours. Reflecting the usual mismatch between the lengths of planetary missions and the scurrying cycles of human lives, the Voyager mission was now on its sixth project manager, Esker Davies. For Davies the situation was more straightforward than it had been for his predecessor in that he would not have to perform any orbital gymnastics.

On August 26, 1981, the first pictures from other moons, notably Hyperion, Dione, and Mimas, started to stream in. In almost every way, the pictures were the exact opposite of the Titan pictures. The images were clear, colorful, and full of detail.

It was not long, however, before *Voyager 2* encountered a problem. Since both *Voyager*s had the ability, as one of their initial design requirements, to hold cameras steady despite their phenomenally fast speeds, they possessed more gears than the best pocket watches (which, by the way, are American railroad grade pocket watches). At the crucial moment of the Saturn encounter—and after all those months of reprogramming—*Voyager 2*'s imaging platform gears locked. Frantic telemetry analysis suggested that the problem might be due to a stray piece of Teflon that had somehow come adrift and was now stuck in a gear. The protocols for troubleshooting were well established: use the ground-based mock-up of the spacecraft to replicate the problem and minutely compare the results with the remote telemetry. Using this technique, Davies determined that it was almost certain that a stray piece of Teflon was the culprit.

When such problems had previously been encountered in the Voyager missions, the engineers had exhibited a certain gung-ho mentality—just give it a good yank. Not this time, however. Davies was well aware of the responsibility he held—the chance to send a spacecraft to Uranus and then Neptune. Despite opposition from some of the imaging scientists, Davies decided not to risk forcing the mechanism. His orders were that the cameras would continue to get whatever pictures they could of Saturn and her moons without jimmying the mechanism that held them. They already had about 40,000 pictures of Saturn and its moons, and it was time to move on.

Perhaps the most spectacular result of the Voyager missions was the information they sent back about Saturn's most renowned feature—its rings. The *Voyagers* showed that Saturn's ring system is made up of seven bands of icy rubble, each with its own intricate structure. The innermost band is only a little more than four thousand miles above Saturn's cloud-tops, and the edge of the outermost band is further away than our moon is from Earth. The *Voyagers* sent back incredible images of previously unknown structures such as kinks and even spokes, as well as entire rings not visible from Earth. As we will see later, these elaborate structures are due to the gravitational effects of nearby satellites.

The rubble of Saturn's rings tells a clear story about their history. The rings are the debris from a moon that either failed to coalesce properly or that broke up—perhaps because it formed too near the Roche limit (the distance at which a planet's gravity field will not permit a moon to form). Given Saturn's enormous size and its strong gravitational pull, its ring system could not, in geological terms, be a particularly long-lasting feature. Given the short period of human existence, we are lucky to be able to see it today. Quite probably the rings were not there when our australopithecine ancestors came down from the trees, and, equally, in ten million years or so when our space-faring descendants return to visit once again the green hills of Earth, the rings of Saturn will no longer exist.

The *Voyager* spacecraft showed that, although this mishmash of rubble is young, it is far from static. Saturn's many moons interact with it in an intricate and mesmerizing dance of orbital mechanics that would make Newton weep with joy. The tiny moon Prometheus is closely linked to the inner edge of Saturn's narrow F ring (which is only slightly more than three hundred miles wide). On the opposite side of the F ring and orbiting on the same radius with the precision of a Miramar fighter pilot is Pandora. Prometheus and Pandora act like cosmic street cleaners, finding any rubble that becomes separated from the F ring and pushing it back into line. Prometheus and Pandora are a type of moon common in the Saturnian system known as "shepherd moons." Shepherding activity is now known to be common to many of the moons of Saturn. A similar task is performed by tiny Atlas, which orbits closer to Saturn near the outer edge of the nine thousand-mile-wide A ring.

Mimas, with its godfather-like scar, orbits the inner periphery of Saturn's outermost ring and street cleans on a scale that can be seen from almost a billion miles away. Mimas maintains the clear region that Giovanni Cassini spotted in the seventeenth century.

Not all of Saturn's moons are street cleaners. Enceladus is tidally kneaded by its neighbors Dione and Tethys so powerfully that it is subjected to continuous volcanism—albeit of the cryogenic variety. Enceladus consequently spews a constant stream of material into the center of the E ring, which makes the ring measurably thicker in that particular orbit.

The diversity of moons in the Saturnian system is easily as great as that of the Jovian system. For example, nine hundred-mile-wide Iapetus is highly reflective on one side and dark on the other. This has a curious effect when viewed from Earth. Since the moon is gravitationally locked so that the same hemisphere always faces Saturn, it only appears when the reflective surface faces the sun, which means that, to Cassini and others who observed it, Iapetus vanishes when on one side of the planet. It was only when the *Voyager* spacecraft arrived that the three hundred-year-old mystery of Iapetus's behavior was finally explained. The most distant satellite of Saturn is Phoebe, which, unlike the other moons, orbits the planet in a clockwise direction and at an angle to the orbits of the others. This orbit suggests that Phoebe is a captured asteroid rather than a native moon of the planet.

Janus and Epimetheus, the two moons closest to Saturn, also behave strangely. They orbit so close to each other—just thirty-one miles apart—that every four years they swap orbits when the inner, quicker one of the pair overtakes the other. The orbit swap is a necessary consequence of the conservation of their orbital momentum. When four years later the inner has once again built up sufficient speed to overtake the outer, the switch happens again. British science fiction author Alastair Reynolds has an even more intriguing explanation for the phenomenon, however. In his epic space opera *Pushing Ice*, Janus turns out to be a spacecraft sent by an alien civilization orbiting the star Spica. When it returns home, it takes a group of pursuing humans with it, thereby seeding the stars with the human race. Other moons of Saturn are similarly locked together. Telesto, Calypso, and Tethys orbit in precisely the same track—one behind the other—while, further out, Helena and Dione follow the same orbit but are spaced precisely sixty degrees apart.

Rhea, orbiting slightly further out still, has the same kind of composition as Dione, Tethys, and Iapetus. It is not gravitationally linked to any other moons and consequently pursues a peaceful, if solitary, existence in the outer reaches of the Saturnian system. The most distant satellite, Hyperion, is a different matter. At two hundred miles in diameter, it appears to have been just too small to complete the process of moon formation by collapsing into a sphere and remains an irregular lump of rock.

The *Voyager* spacecraft showed that many of Saturn's moons have at least some of the essential ingredients of life: carbon, water, and perhaps oxygen and hydrogen. Only on Titan, however, are these ingredients most tantalizingly combined, and, for this reason, sixteen years after the *Voyager 2* flyby, a mission was sent out not just to view Titan but to land a probe on its surface as well.

## BACK TO THE FUTURE

The origins of the Cassini-Huygens mission began only a year after *Voyager 2* passed through the Saturnian system. On the strength of the images *Voyager 2* returned, a small group of European scientists suggested that the time had come to dispatch both an orbiter and a lander to the ringed planet. The eye-watering costs of such a mission motivated NASA and the European Space Agency to collaborate. By 1985 a joint NASA/ESA feasibility study of the mission concluded that the mission could be achieved; NASA would build the orbiter, and the ESA, the lander.

This type of joint enterprise was something of a departure for NASA, which for decades had operated as if not just our world but all the worlds of the solar system were its own personal oyster. As had happened many times throughout NASA's history, the agency's desire to work with the ESA was in part motivated by the Soviet Union. Frustrated with NASA's isolationist dominance, the European space science community had started collaborating with the Soviets. It was enough to give NASA pause and to crystallize the thought that collaboration with the Europeans might, after all, be a good thing.

The close relationship that eventually developed between NASA and the ESA saved the Cassini-Huygens mission on more than one occasion. At various times through the years, Congress looked for ways to cut the mission's funding but, wary of offending the Europeans, backed off. By 1994 the project was moving ahead smoothly, with a projected launch date in 1997.

It had become clear early in the project that the *Cassini-Huygens* spacecraft would be the most elaborate unmanned probe ever built. The *Cassini* orbiter weighed more than 2,000 kilograms, while the *Huygens* probe alone weighed in at 350 kilograms. At launch the *Cassini* spacecraft stood almost seven meters tall and was more than four meters wide. The entire spacecraft cost in excess of $3 billion. It was a bold mission, and NASA and the ESA were spending almost an eighth of the Apollo moon landing budget to get to the other end of the solar system. *Cassini*'s instrumentation included a sophisticated synthetic aperture radar mapper like that used so successfully to unveil the secrets of Venus, several spectrometers for analyzing the chemistry of the Saturnian system, and instruments for measuring the magnetic fields of Saturn and her moons. The *Huygens* probe was equipped with sophisticated sensing apparatus to explore Titan's atmosphere during its descent, as well as a cunningly contrived experiment (built by John Zarnecki's team, then based at the University of Kent) to measure the nature of the surface upon landing. Like the *Voyager* and *Pioneer* spacecraft, the *Cassini-Huygens* probe was powered by a plutonium RTG. It was the same design as that used on the *Galileo* spacecraft to Jupiter, which had already proved its reliability, and it carried over thirty kilograms of plutonium dioxide—the largest quantity of radioactive fuel ever to be launched into space.

Despite vociferous protests about the use of the plutonium-powered RTG, *Cassini-Huygens* successfully launched from Cape

Canaveral on October 15, 1997. The mission profile called for a number of gravity assist flybys, passing Venus twice, Earth once, and then sweeping past Jupiter. With this varied itinerary, *Cassini-Huygens* not only provided new information about Saturn but also several other planets, asteroids, and moons en route. Ironically, *Cassini-Huygens*'s Earth flyby took it around the dark side of the moon, allowing it to make a series of detailed calibration images. A body that had once been the target for the most colossal expenditure in the history of humanity was now reduced to the status of a testcard.

*Cassini-Huygens* reached Jupiter in December 2000 and took almost 26,000 images during the course of a flyby that lasted several months. With its state-of-the art cameras, it produced a detailed color portrait of Jupiter showing features as small as forty miles across. Scientifically, too, the flyby was a resounding success, for it answered a long-standing question about the nature of the Jovian atmosphere—were the light zones areas of atmospheric upwelling? The *Cassini-Huygens*'s imagery showed that individual storm cells do exist but that they occur as small, localized patches of white gas in Jupiter's dark belts. *Cassini-Huygens* also discovered a region of swirling dark haze near the north pole that was about the same size as Jupiter's most famous atmospheric feature, the Great Red Spot.

*Cassini-Huygens* also revisited an experiment that had made Mercury famous almost a century before when it was used to test Einstein's then new theory of general relativity. Firing radio signals home that passed close to the sun, *Cassini-Huygens* showed that, as predicted, the signals took longer to reach Earth than those that did not pass close to Sol. This result accords precisely with Einstein's theory, which predicts that space-time is curved in the presence of massive objects and that electromagnetic radiation traveling through such regions will take longer to reach its destination.

*Cassini-Huygens* arrived in Saturnian space on July 1, 2004, slipping gracefully through the gap between the F and G rings before taking up orbit around the planet, the first spacecraft ever to do so.

*Cassini-Huygens* did not waste any time and made the first flyby of Titan the next day. Unlike previous missions, the spacecraft was properly equipped to investigate that strange, murky moon. From a distance of 211,000 miles and using special imaging filters, *Cassini-Huygens* showed that the clouds over the south pole are composed of methane, and the instruments hinted at some surface topography beneath Titan's all-enshrouding haze. In October the spacecraft returned for the first of what were to be many very close flybys of Titan, skimming the clouds only a few hundred miles above the moon's surface. Using its synthetic aperture radar, *Cassini-Huygens* confirmed the presence of a complex topography, although there was as yet no obvious evidence of lakes of liquid hydrocarbon, as some had speculated.

Meanwhile, preparations were under way for one of the highlights of the mission, the release of the *Huygens* probe. The orbiter released the probe on Christmas Day 2004. There was almost a sense of déjà vu for me, because once again I was at home for Christmas, with my family in West Oxfordshire waiting for the outcome of a space mission. It had been only a year since I had padded downstairs and waited in vain for the distant voice of *Beagle 2*. This was to be a very different occasion, however. Shortly before lunch, while my father and I were contemplating the prospect of aperitif, I received a text message from my colleague Bob Spicer of the Open University telling me that he had just heard from John Zarnecki, who was at mission control in Pasadena. The release of the *Huygens* probe had gone according to plan, and all systems were go for a Titan landing.

From Saturn, to Pasadena in California, to Bob's home north of Oxford, to my own home in a small West Oxfordshire village, we were tiny humans celebrating the birth of a two thousand-year-old carpenter while waiting for a message from the edge of the solar system. We were a billion miles from the action, but our ability to talk and interact kept us all engaged—compulsive communicators exploring the solar system from beside our Christmas trees. The juxtaposition of scales—from the colossal to the homely—was at once staggering and uplifting. Is there anything we cannot do if we simply put aside our differences and decide to work together?

Amazingly, it would take over three weeks for the probe to finally touch down on Titan's surface, but most of the action would occur in the last four hours before landing, when the probe entered the atmosphere. Fifteen minutes before slicing through the moon's legendary atmosphere, *Huygens*'s automatic timer activated the probe's systems. The probe's batteries were designed to last three hours, powering it for the two-and-a-half-hour descent through the soupy atmosphere and for at least thirty minutes on the surface—assuming the probe, as everyone hoped, landed softly. The probe's radio link was switched on early in the descent, and the orbiter began collecting data. At the end of the descent, the orbiter's high-gain antenna turned toward Earth, and the data downloaded.

*Huygens*'s transmitter was only ten watts—about the same power as a mobile phone. Incredibly, though, very large radio telescopes on Earth could pick up the signal directly. The Greenbank telescope in West Virginia was the first to detect *Huygens*'s tiny shout from deep space. The signal, being so weak, had to be collected and then put together as a backup in case the data dump from *Cassini* failed.

*Cassini*'s first close flyby of Titan had suggested that *Huygens* might land near the shoreline of a liquid ocean. As the telemetry came in from the descending probe, it showed a mist-shrouded coastline criss-crossed by massive drainage channels that fed a nearby dried-up lakebed that might once have held liquid methane. Chunks of ice were scattered over an orange-hued coastal plain, above which wisps of methane gas tattered in a gentle breeze. The superficial similarity to Earth was extraordinary. The surface of Titan itself, as revealed by John Zarnecki's penetrometer, appeared to be of a clay-like material underlaid by a region of looser consistency. One joker named it a "crème Brulee" world.

By any standards, the landing on Titan was a stunning success. Yet the mission to Saturn had not been without its tensions. Long after the spacecraft had departed Earth, two European engineers, Claudio Sollazzo and Boris Smeds, discovered that *Cassini*'s receiver had a design flaw that would potentially prevent any data receiver by the *Huygens* probe from being returned to Earth. Smeds persuaded his colleagues to perform tests on the transmitter while *Cassini* was en route, and to their horror, they discovered that he was right. The probe's radio signal during the Titan descent would experience a Doppler shift from the probe's own acceleration, and although the orbiter's software had been configured to account for this shift, the firmware had not. It was impossible to reprogram the firmware (the relevant microchips were already millions of miles away and receding fast), so the decision was taken to re-orient *Huygens*'s trajectory so that it would move perpendicular with respect to the orbiter during the descent, thereby curing the problem.

The fix worked perfectly, but another problem soon surfaced—this time during the orbiter's descent. *Huygens* used two redundant radio channels to transmit data back to *Cassini*. Long before the mission had departed Earth, the principal investigators were given a choice—split the same data over both channels (the classic double-redundancy approach) or rely on one each and get twice as much data. There were six experiments on *Huygens*, and the designers of five of them had gone for the greedy approach—they would try to squeeze as much data out of the descending probe as they could. Because of a problem with one of the channels, nearly all of the wind speed and direction data were lost, and only half of the anticipated seven hundred pictures were received. Only John Zarnecki's group—the Open University team who designed the surface science package—had gone for the redundancy approach. Zarnecki had been in the space industry for almost thirty years, and he was too old a hand to put all his eggs into one interplanetary basket. Zarnecki's team lost no data.

In the end the problem with the channel was traced to a simple software oversight. Somewhere in the labyrinthine bowels of the

ESA's quality-control procedures, someone had failed to add one of the two commands necessary to tell the channel A receiver on *Cassini* to switch on when told to by *Huygens*. Whoops.

Although mistakes were made, other aspects of the mission broke new and revolutionary ground. Imaging scientists made the raw images sent back by the spacecraft available to the public via the Internet. Many thousands of small, low-contrast images had to be assembled into mosaics and panoramas of the landing region, a vast and time-consuming process for the relatively small Huygens team. Only hours after the first images of *Huygens*'s landing site had been released by the team at Darmstadt, however, a space enthusiast named Christian Waldvogel posted a panorama of the region on the Web. Another enthusiast, René Pascal, developed a method to remove camera artifacts from the images. It was a fitting end for a mission that began as a collaboration between two giant, international space agencies and a heartwarming tribute to the spirit of the online age.

## THE LOST WORLD

We went to Titan because it seemed the world most similar to the Earth when our world was new—soon after the formation of the solar system, when our planet was awash with organic chemicals and long before the rise of oxygen. We wanted to explore the strange chemistry that ultimately gave rise to our own world and our own species in a place where that chemistry might still be happening. Of course, the analogies between Titan and early Earth are not perfect—for starters, Titan is much colder than the early Earth was. Still, Titan held the potential to serve as a comparison to the lost world of Earth.

Before *Huygens* landed on Titan, it was another mysterious object in the solar system. Larger than Mercury, the main thing known about it was that it had a dense atmosphere. It was the largest unexplored surface in the solar system. Now, thanks to the Cassini-Huygens mission, we know a little about Titan's secrets. Titan, like Earth and Mars, has channels cut by the action of flowing liquid. At fifty to a hundred meters deep, these channels suggest that methane rain does indeed fall on this remote, forbidding moon, the only other world in the solar system—apart from our own—where atmospheric precipitation is known. There is probably not quite enough atmospheric methane to cause showers or to feed lakes, and no large methane oceans have been found, but the possibility of an occasional methane monsoon exists.

*Huygens* confirmed that Titan's rich atmosphere is derived from the breakup of methane in the upper atmosphere by ultraviolet light from the sun, which then recombines into other large organic molecules like hydrocarbons. These gasoline fumes are what create Titan's characteristic haze. Carbon isotope measurements strongly suggest that Titan's methane is not produced biologically. Instead it may be generated deep beneath Titan's surface by geological processes that we have only begun to understand. There is a thirty-km-wide topographic upland on Titan's surface that could be a cryovolcano, a gateway to this strange, new, and incredibly distant geological underworld. Surface geological features—like gorges and canyons—mutely testify to the occurrence of some kind of low-temperature volcanism.

*Huygens* touched down on a surface with the consistency of wet sand, the spacecraft's relative warmth sending a plume of methane and hydrocarbons into the moon's sluggish atmosphere. Carl Sagan, writing twenty years ago in *Cosmos*, said "there is no strong evidence either for or against life on Titan. It is merely possible. We are unlikely to determine the answer to this question without landing instrumented space vehicles on the Titanian surface." Although Sagan's wish for landing a vehicle on Titan has come true, it is still not known whether complex organic molecules can be combined there to make the amino and nucleic acids with which we started our own long journey to sentience.

Yet Titan remains one of the most promising places in the solar system to find such molecules because of its still-functioning photochemistry, a chemistry that must have been important on our prebiotic Earth but that has since been superseded by the more violent enzyme-catalyzed reactions of our turbo-charged bodies. Consequently, a return visit to Titan is very much on the cards. NASA's Outer Planets Assessment Group, which sets the scientific goals for missions to the solar system's outskirts, ranks Titan only behind Europa as the most-favored candidate for a closer look. Europa heads that list because it remains the most likely candidate to have extant life.

Meanwhile, John Zarnecki thinks that the next probe to Titan will carry a rover, but one designed to take advantage of Titan's most notable characteristic: its thick atmosphere. Why struggle over a complex and largely unknown terrain when the thick atmosphere and low gravity cry out for a dirigible? The many complex technical difficulties that had to be overcome to make the Mars Exploration Rovers the stunning success they are can be completely bypassed on Titan by something as simple as a balloon.

To date, perhaps the most impressive thing about the *Cassini-Huygens* mission is that it got there at all. We achieved something almost out of the realm of the imagination; we built a spacecraft and landed it on the surface of an unimaginably faraway world. The *Huygens* probe is the most distantly controlled landing yet made by the human race. It will not be the last.

# 9

# The Harmonies of the Ice Giants: Uranus and Neptune

Because the gap between the arts and sciences seems almost unbridgeable today, it is worth remembering that the greatest astronomical discovery of the eighteenth century was made by a man whose first love was music. William Herschel was born in 1738 and took almost three decades to develop the interest for which he would be known to history—astronomy. His curiosity about astronomy was kindled after reading Robert Smith's *Compleat System of Optiks* during the winter of 1772, and Herschel decided that he wanted to see the planets for himself. He experimented for a time with "off the shelf" lenses but soon found them inadequate for his purposes, so he started making his own lenses and mirrors. By 1776 the quality of his homemade telescopes had surpassed even those of the Royal Observatory at Greenwich.

It was 1781 when he made the discovery that made him famous. Herschel was engaged in a survey of eighth-magnitude (dim) stars when he came across one that he initially thought was a comet. Because the disk increased in size with increasing magnification, however, Herschel began to suspect the object might be a planet. The next step was to calculate its orbit—the orbits of comets are highly elliptical. This task was duly done, and the result was earth shattering—the orbit was broadly circular, the diagnostic criterion of a planet. That planet became known as Uranus.

It is difficult now to appreciate just how important this discovery was to eighteenth-century science. Until Herschel's finding, no new planets had been detected since the time of the ancients. But even more significant, Herschel's discovery was a spectacular confirmation of the Titius-Bode law. Uranus was exactly 19.6 AU from the sun, precisely where the law predicted a planet should lie. Astronomy's new status as a predictive rather than a descriptive science was resoundingly confirmed. It was Bode himself who proposed the name of the new planet: Uranus, the father of Saturn and grandfather to Jupiter.

Uranus is unique in that it spins on its back; that is, its axis of revolution is parallel to the plane of the ecliptic and not at right angles to it as with all the other planets of the solar system. The result is the curious condition whereby Uranus's poles receive more solar energy than its equator. Despite this orientation, however, Uranus's equator is still warmer than its poles through an as yet unknown mechanism. Because of the planet's laid-back orientation, it is difficult to ascribe a north or south pole. Uranus is thus qualitatively different from the other planets of the solar system.

## DEEP BLUE

*Voyager 2* slipped quietly and without fanfare into the Uranian system in late January 1986. It had been fifty-three months and almost another billion miles since the small spacecraft had left Saturn. In a way, *Voyager 2* was now being asked to do what *Voyager 1* had done at Titan half a decade before: some stunt flying. The reason was related to the strange topography of the Uranian system. Because Uranus rotates tipped on its side, its rings are effectively upright, and a spacecraft must somehow pass through this rotating nebulous wall. At JPL the trajectory planners were pretty sure the planet had at least nine faint rings, every one of which was orientated almost vertically. An additional hazard was that Uranus is circled by at least five comparatively large moons, Miranda, Ariel, Umbriel, Titania, and Oberon, which range in size from three hundred and to almost one thousand miles in diameter. These, too, orbit vertically and loom as rather dangerous objects in the Uranian system. And those were just the known objects to avoid; *Voyager 2*'s planners had no way of knowing how many other rings and satellites might be lurking around Uranus.

To forge a way through this interplanetary rubble and still head on to Neptune would require some navigational compromises. The mission planners knew that, to maintain the necessary safety margins, they would be able to make a close flyby of only the smallest of the known Uranian moons—Miranda. This trajectory would keep *Voyager 2* a safe distance from the planet as well as any other satellites, but the tradeoff would be a lack of close-up pictures. It was a hint of the way things had changed at JPL over the years. As the mission had extended itself further and further and the possibility of getting to Uranus and Neptune increased, the mission scientists felt less and less inclined to take chances. Who could blame them? After a decade in deep space, *Voyager 2* was the only game in town when it came to understanding the harmonies of the ice giants, and replacing the spacecraft if they lost it was simply not an option. It wasn't as though you could pull into the nearest repair shop if something broke. And besides, all of this was bonus. There were those around JPL who remembered only too well that, for the Nixon government, neither Uranus nor Neptune was even on the agenda.

*Voyager 2* had spent the four years since leaving Saturn whistling through the outer solar system with most of its systems powered down. Mission control would occasionally check that things still functioned on the small spacecraft, but otherwise the overwhelming requirement was to save power. Uranus was still very much a mystery world. Only some basic facts were known: it is the third largest planet in the solar system with a diameter of 32,000 miles, and it orbits the sun at a distance of about 2.8 billion kilometers (1.7 billion miles), completing one orbit every eighty-four years. But even such basic facts as the length of the Uranian day were unknown before *Voyager 2* arrived.

As the spacecraft arrived in Uranian space on January 24, 1986, the little machine was whirring and clicking once again, taking images and collecting data that would take over 2.5 hours to reach Earth. But when the data did arrive, it was immediately obvious that the pictures had been worth the wait. One of the first images the waiting scientists received was of the surface of Miranda, and what a surface! It was unlike anything anybody had ever seen on a planet or a moon, a broken, fractured, and tormented mess. Smooth plains lay cheek-by-jowl with ragged scarps and shattered mountain ranges. It was the most varied terrain in the solar system.

The pictures *Voyager 2* returned showed that Uranus is enigmatic, even more so than Titan. Whereas Titan is a featureless orange orb, Uranus has a soft blue color. What they both have in common is a complete lack of any distinguishing features. There are no clouds, swirls, or spots to disturb the monotony of Uranus. Chemical sensors aboard the spacecraft showed that the planet's atmosphere is 83 percent hydrogen, 15 percent helium, with the rest being methane and a handful of miscellaneous hydrocarbons. The craft's magnetometer showed that the planet's magnetic field has a curious corkscrew shape unlike any other planet in the solar system. Scientists believed that this shape was a result of the planet's curious sideways orientation.

Uranus's rings and satellites are as impressive as the planet's curious laid-back orientation. The rings are nebulous formations, fainter by far than Saturn's, and *Voyager 2* uncovered two additional, unknown rings, bringing the total to eleven. Uranus's rings are most likely the result of the same processes that formed Saturn's rings: remnants of a moon that broke up when it strayed inside the Roche limit. Also like Saturn's, Uranus's rings are probably a temporary feature and, within a few million years, will vanish.

The JPL people were fairly sure that Uranus would have more moons than could be seen from Earth, and to be sure that *Voyager 2* would discover as many as possible, a scientist named Steve Synnott developed new satellite-hunting software. If an imaging team scientist found a suspicious point of light in a picture, the image could be loaded into a computer that would scan multiple images and alert the scientists if the point moved. If it did, the computer would calculate the trajectory of the object's orbit.

Synnott's satellite-hunting software turned out to be invaluable. When *Voyager 2* pointed its cameras at the Uranian system, a host of other moons besides the five well-known and relatively large ones immediately sprang into view. Ten previously unknown satellites circled the planet, some measuring up to 1,500 miles in diameter. All of them looked exceedingly rich in carbon, which immediately raised the prospect—however faint—of life in the Uranian system. The orbits of the moons are orderly, which suggests that they are native features of the Uranian system and not captured asteroids or Kuiper Belt objects. The first nine new moons were named Cordelia, Ophelia, Bianca, Cressida, Desdemona, Juliet, Portia, Rosalind, and Puck, all reflecting an infatuation with Shakespeare. The tenth moon was named Belinda after a character in Alexander Pope's epic poem, *The Rape of the Lock*. Cordelia and Ophelia act as shepherd moons for Uranus's bright Epsilon ring, while the other small moons orbit alone, having little gravitational influence either on the planet or each other.

The mission scientists found Uranus's large moons full of surprises. Ariel, while measuring only about seven hundred miles across, shows evidence of resurfacing—something no one expected in such deep space. Ariel's craters look as though they were flooded by some molten material at some point in the remote past, but what the material could be in that extreme region is anybody's guess; theories range from rock to methane. Umbriel, although about the same size as Ariel, is the darkest of the large Uranian moons. With its surface pockmarked in ancient craters, Umbriel appears to have changed very little since its formation. Umbriel is a relatively static world, one that has not experienced any of the deep upwellings that have smoothed the face of Ariel.

Oberon is the most distant orbiter of Uranus and, at slightly less than a thousand miles in diameter, is similar to ancient Umbriel, with a dark, heavily cratered surface that is probably as old as the solar system. *Voyager 2* spotted a four-mile-high mountain on Oberon's surface. Scaling for planetary size, such a mountain would be thirty miles high on Earth. Titania is about the same size as Oberon but resembles Ariel with its white, cleanly polished surface. Once again the visit to Uranus was demonstrating—in the most graphic way possible—the diversity of the solar system.

The most extraordinary moon of the Uranian system is the one *Voyager 2* first saw, Miranda. The small moon measures less than three hundred miles across and is the closest to Uranus itself. As better images came into JPL from *Voyager 2*, the imaging scientists made out features as small as only a few hundred feet across. The diversity of the terrain confirmed that Miranda was apparently torn apart several times in the geological past, only to reassemble itself again. That this abuse happened on more than one occasion suggests that chance meteorite impacts were not the culprits and that Miranda's tortuous terrain has been caused by some heavy-duty tidal kneading.

*Voyager 2*'s visit to the Uranian system was fascinating but fleeting, because on January 25, 1986, less than twenty-four full hours after it arrived, the craft left Uranus behind. *Voyager 2* skimmed Uranus's cloud tops at an altitude of a mere (in deep-space reckoning) 50,600 miles and, during its fleeting encounter, added immeasurably to what we know about the planet. But even after the probe left Uranian space, the scientists continued to learn about the planet as they pored over *Voyager 2*'s data.

The planet's deep blue color is due to its methane-rich upper atmosphere, which absorbs red light. It is now known that this atmosphere has bands and swirls of color that vary as a result of Uranus's orbit around the sun. Incredibly for a planet that is practically at the edge of the solar system, it seems that Uranus has seasons! Adding to Uranus's general air of mystery is the question of what caused it to rotate on its back. The best guess so far is that, at some point in the distant past, Uranus was hit by a passing comet or asteroid that failed to destroy the planet but was sufficiently large to knock its axis of rotation out of kilter by ninety degrees.

Besides the detail about Uranus's atmosphere and the nature of its moons, *Voyager 2* amplified our basic knowledge of the planet, down to clocking the length of a Uranian day as seventeen hours, fourteen minutes. We now know that Uranus has a magnetic field and that its intensity is roughly comparable to that of Earth's. The Uranian magnetic field, however, is much more variable because of its large offset from the center of the planet, probably a result of the planet's sideways rotation.

Uranus's radiation belts were found to have an intensity similar to Saturn's. The intensity is such that any methane in the inner moons or rings is chemically altered quickly so that its appearance darkens within about 100,000 years. This may be the cause of the rings' and moons' characteristically uniform dark color. Uranus's layer of high haze around its sunlit pole radiates large amounts of ultraviolet light, a phenomenon the JPL team dubbed "dayglow." But this is not a high-temperature phenomenon, for the average temperature on Uranus is about 60 kelvins (–350 degrees F), and the temperature is very uniform both across the planet's surface as well as vertically through its atmosphere.

Finally, *Voyager 2* showed that, although Uranus has a ring system probably formed by the destruction of a moon, in other ways it is fundamentally different from the ring systems of Jupiter and Saturn. Uranus's ring system may be relatively young and did not form at the same time as the planet.

## THE PLANET ON THE EDGE OF FOREVER

Not long after the discovery of Uranus, Alexis Bouvard, a shepherd boy from Chamonix in France who ended up having a glittering career in astronomy, started to compile information on its orbit. Uranus had been spotted several times before Herschel declared it a planet, and Bouvard synthesized these "pre-discovery" observations to accurately calculate the orbit. What he found stunned him and the rest of the eighteenth-century scientific establishment: Uranus's orbit was irregular. There were only two possible explanations for this irregularity: either Newton's laws were fundamentally flawed, or there was another planet beyond Uranus that was influencing it. It was impossible to believe that Newton's laws, which so spectacularly explained the motions of the heavens, could be flawed, so the hunt was on to find the new planet. They would not be looking completely in the dark, however; as with Uranus, the Titius-Bode law predicted that the new planet would be found about 38 AU from the sun.

The story of Neptune's discovery is one of the most extraordinary in the history of planetary astronomy—and is a cautionary tale for science as a whole. It revolves around the slight and diffident figure of John Couch Adams. Adams was a brilliant mathematician who went to Cambridge in 1839 and, three years later, graduated with stunningly high marks in mathematics. As early as 1841, Adams started to take an interest in the problem of Uranus's orbit. He knew the solution would have to be mathematical, and he spent the next four years working on the problem. By September of 1845, Adams had calculated the orbit and position of the new planet. On October 1, 1845, en route to his family home in Cornwall, he left the coordinates for George Airy, astronomer royal at the Greenwich Observatory. He requested that Airy check that particular segment of sky on a prescribed date, and he continued home confident that the new planet would be found. Airy, however, did not follow up on Adams's request. At the same time, and unknown to the Englishmen, a parallel study led by the Frenchman Urbain Leverrier was being conducted. By the summer of 1846, Leverrier had also calculated the position of Neptune's orbit, and he accordingly informed Airy. But Airy neither mentioned Adams's work to Leverrier or Leverrier's work to Adams. Disenchanted by Airy's lack of interest, in September 1846 Leverrier asked the German Johann Galle to start a search for the planet. Within the first hour of searching, Galle found an eighth-magnitude star exactly where it was predicted to be; Neptune had been found.

The repercussions from Airy's failure to act on Adams's request to search for Neptune were considerable. There was a terrific row between the British and French scientific establishment as to who should be given priority for the discovery. Adams had clearly made the calculations first, yet he had not published them and Airy had not made Adams's work known. Adams behaved like a gentleman and, avoiding argument, generously gave Leverrier credit for the discovery. Airy himself, a sarcastic and rigid martinet, eventually became president of the Royal Society, having neatly sidestepped the blame that should have been his for failing to act on Adams's advice. Even today the acrimony surrounding the discovery of Neptune occasionally resurfaces in another venerable British institution—the hallowed halls of the Royal Astronomical Society.

Strangely enough, Neptune was certainly observed by Galileo himself in 1613, but the sage of Florence mistakenly thought it was a star. If he had begun his observations a couple of days earlier, he could have observed Neptune's movement, and the planet might have been discovered two centuries earlier! Galileo's efforts were frustrated, however, by that most mundane of Earthly phenomena—cloudy skies. On such tiny details does the progress of science hinge.

Neptune is the second of the ice giants and arguably the outermost planet of the solar system. For twenty of its several hundred years' orbit, it crosses Pluto's orbit and becomes the furthest out even of the "conventional" planets. However, as we shall soon see, there are other good reasons for regarding Neptune as the last planet in the solar system.

Geologically Neptune's composition is probably similar to that of Uranus, with a similar suite of rocks and super-cold gases such as methane. However, unlike Uranus but in a similar manner to Jupiter and Saturn, Neptune does have an interior heat source and, therefore, radiates more heat than it receives from the sun. Like the other gaseous planets, Neptune has an active atmosphere, with rapid winds running in jet streams as well as large-scale storms and vortices. Its Great Dark Spot, the diameter of the Earth but about half the size of Jupiter's better-known Great Red Spot, was observed by the only spacecraft ever to visit Neptune, *Voyager 2*. Since that epochal encounter in 1989, however, observations from the Hubble Space Telescope have shown that the Great Dark Spot has disappeared. Clearly Neptune's atmosphere is very active, an idea supported by the fact that Neptune's winds are now known to be the fastest of any planet in the solar system, routinely reaching speeds of up to two thousand kilometers an hour.

Like Uranus, Saturn, and Jupiter, Neptune has a well-developed ring system and several satellites. The largest moon is Triton, discovered in 1846 by the astronomer William Lassell. All of Neptune's satellites have very strange orbits; the hypothesis is that Triton, a moon larger than Pluto, is a captured ice body from the Kuiper Belt and that its arrival destroyed the orbital symmetry of the rest of Neptune's moons.

*Voyager 2*'s encounter with Neptune was the most problematic of all its planetary visits for the JPL mission planners. At such extreme distances from the sun, light levels are incredibly low and would tax even *Voyager 2*'s cunningly contrived, swiveling, and temperamental Teflon-geared camera platform. The spacecraft would be moving at 35,000 miles per hour, which compounded the difficulty of getting good photos. The JPL imaging scientists decided to compensate for the high speed and poor light by instructing *Voyager 2*'s cameras to take a series of close-up images that would be stitched together into panoramas.

Another, even bigger problem with the Neptune encounter had to be solved, however. Spectroscopic studies from Earth suggested that Neptune's moon Triton, like Titan, might be an interplanetary hotbed of organic chemicals such as ammonia, carbon, nitrogen, hydrogen, and oxygen. That alone made it worth a visit, but added to that was Triton's weird orbit. Triton has a retrograde orbit—that is, it orbits in the opposite direction to its primary—the only satellite in the solar system to do so. Such an outré orbit could easily churn up some serious tidal kneading in Triton's interior and warm things up enough for those organic chemicals to start doing interesting things—like combining themselves into organic molecules, perhaps into a prebiotic soup.

To get close enough to Triton to take a look would mean skimming *Voyager 2* above the north pole of Neptune at an altitude of just 3,000 miles. This approach would hand-brake turn the spacecraft so that it could head downward out of the plane of the ecliptic—in so doing passing Triton. The problem was that Neptune is almost 3 billion miles away and 30,000 miles in diameter at the equator. Directing a spacecraft through a 3,000-mile maneuver at such distances and with such fine tolerances would be far from easy. It was *Voyager 1*'s Titan encounter at Saturn all over again. An added hazard was that Neptune was known to have a thick atmosphere, and very little was known about its depth. Would 3,000 miles be sufficient clearance? Nobody knew for sure, but they did know that they would need the best possible communications with the spacecraft.

The most advanced radio telescope in the world was the Very Large Array in New Mexico. It was even better than JPL's own Deep Space Tracking Network. The only problem with the Very Large Array, as far as the Voyager team was concerned, was that it was in Very Great Demand by the rest of the world's astronomical community, and the JPL scientists needed three months of uninterrupted access to the facility while they prepared for and then executed the Triton encounter. They needed some serious leverage for this to happen. Fortunately, they found it. The Very Large Array's software was not up to snuff; it did not fully exploit the capabilities of the telescope's state-of-the-art hardware. JPL offered to upgrade the software for free, gratis, in return for their ninety days of exclusive access. When it came to computing and astronomical expertise, JPL's capabilities were unsurpassed. To no one's surprise, the directors of the Very Large Array accepted the offer.

*Voyager 2* reached Neptune on August 25, 1989. As the spacecraft approached, the first noticeable thing was that the planet, despite its staggering distance from the sun, was a hotbed of activity. *Voyager 2* sent back to the listening ears of the Very Large Array the incredible news that Neptune was radiating twice as much heat as it absorbed from the sun. There could be only one explanation: Neptune must have its own radioactive power plant like the planets of the inner solar system. The possession of such a serious energy source gives Neptune some pretty extreme weather conditions. At the top of the atmosphere, it is as cold as you would expect for a planet that orbits the edge of the solar system—almost –400 degrees F. But near the surface, the temperature is over 900 degrees F. This difference produces some pretty extreme wind speeds. Around Neptune's equator, winds blow in excess of 1,200 miles per hour. That is almost twice the speed of sound on Earth. Furthermore, the royal blue face of the planet has distinct patches of high-altitude cloud that look very like the icy cirrus formations of Earth. The southern hemisphere has its own significant atmospheric feature, a Great Dark Spot as big as the Earth, around whose circumference westerly winds blow at nearly seven hundred miles per hour. Those who thought Neptune would be a silent, deep-frozen outpost of the solar system could not have been more wrong. The place heaves with meteorological malice.

And Neptune has rings. Until *Voyager 2*'s flyby, scientists believed that Neptune's rings were incomplete, not even girdling the planet's circumference. This belief sprang from Earth-based observations of stars winking on and off when viewed close to the edge of Neptune, a phenomenon most interpreted as caused by incomplete rings intermittently obscuring the stars. But the story turned out to be more complex. *Voyager 2* showed that Neptune has four concentric rings but that they are not the neat affairs found around Saturn and Uranus. Rather, Neptune's rings are rough and unfinished, with areas dense with rubble (and therefore opaque) next to regions where the density of material is almost nebulous. Starting from the outermost ring, the rings have been designated Adams, Plateau, Leverrier, and Galle.

As at Uranus, *Voyager 2* swiftly added to the catalog of moons known to be orbiting Neptune. Previously only giant Triton and tiny Nereid had been known, but *Voyager 2* discovered another six objects orbiting the planet at altitudes of between 30,000 and 73,000 miles. The International Astronomical Union named the new moons Naiad, Thalassa, Despina, Galatea, Larissa, and Proteus after mythological water deities. Unlike massive Triton, none looked like objects that had been captured from outside the Neptunian system but rather like fragments from a moon that had been formed from the same cloud of accretionary debris that had formed Neptune and that had broken apart under the influence of collisions. Proteus, the outermost of the newly detected moons, is similar to Saturn's Hyperion in that it is very nearly—but not quite—large enough to collapse in upon itself to form a sphere.

*Voyager 2* confirmed several basic facts about Neptune; for example, it orbits the sun every 165 years, and the length of a Neptunian day is 16 hours, 6.7 minutes. Neptune's magnetic field, like that of Uranus, turned out to be highly tilted—forty-seven degrees from its rotational axis and offset at least 8,500 miles from the physical center of the planet. The skewed magnetic field on Neptune set scientists wondering if they had been mistaken about Uranus. Are offset fields merely a consequence of magnetic materials flowing in the active centers of both Uranus and Neptune?

JPL scientists are still not sure about the forces acting on the ice giants' magnetic fields, but they did soon realize that they were definitely wrong about Triton's life-originating possibilities. *Voyager 2*'s thermal sensors found that Triton, unlike Neptune, is cold. At about –391 degrees F, it is the coldest body known in the solar system. Far from being a hotbed of potential prebiosis, Triton is merely a deep freeze for the chemicals that could initiate life but simply won't without some kind of energy input. Triton's surface, though, turned out to be highly reflective, which suggests that, at some point in its distant geological past, a lot of internal energy resurfaced on the planet. Triton even shows evidence of oceans, which could have been composed of prebiotic precursors such as ammonia and common water. In fact, tidal heating could even have caused Triton's surface to have been liquid for as long as 1 billion years after its capture by Neptune.

If Triton could somehow be picked up and moved closer to the center of the solar system, its rich pickings of organic chemicals might be inclined to show signs of life. But as it stands, Triton is covered in a world-girdling crust of solid nitrogen, methane, carbon monoxide, and carbon dioxide. There is no action there; Triton is the world that never happened.

Strangely, though, Triton is not completely without surface activity. At its closest approach, *Voyager 2* imaged features that looked like volcanic geysers. Here and there above the surface, dark plumes rose into the sky, while grayish stains elsewhere on the eternal ice testified to the presence of such plumes in the past. For a while, the JPL scientists wondered if Triton was not as geologically dormant as its ice-encrusted surface suggests, but they eventually concluded that the marks are actually caused by the largest feature of the solar system—the sun. Even at that phenomenal distance, the tiny radiation reaching Triton from the sun is leaving its mark. The weak solar radiation penetrates the transparent nitrogen ice-crust of Triton's surface, warming the dark, absorbent, carbon-rich layers underneath. Eventually, in localized regions, sufficient energy is absorbed to sublimate the nitrogen ice to gas, which explodes through cracks in the ice and flashes into the Tritonian sky as geysers of soot.

The final sad note about Triton, though, is that its long, lifeless history is unlikely to have a happy ending. Deep-space salvage tugs of the faraway future will never tow Triton back to the life-giving warmth of the inner solar system to kick-start life on its cold surface, because Triton will no longer exist. Its retrograde orbit forces it to eternally fight Neptune's gravitational field, and a gentleman's accommodation between the two bodies will never be found. Given Neptune's size, Triton will lose the fight eventually and fall into Neptune's gravity-well, breaking up and forming, in all probability, another ring. As all gamblers know, the house always wins.

On August 25, 1989, JPL received a signal confirming that *Voyager 2* had passed the last of Neptune's moons and was dropping out of the bottom of the solar system to begin its eternal fall into the vastnesses of space between the stars. In slightly less than 300,000 years, it will make its first approach to another star—the hot, bright furnace of Sirius.

# 10

# Termination Shock: Pluto and the Kuiper Belt

Nothing beats the thrill of a rocket launch. In 2000, the final year of the millennium, my family and I found ourselves staying once again at a favorite hotel in the Floridian coastal resort of Cocoa Beach. We had arrived the day before on the strength of a previous visit that we had enjoyed immensely. On that occasion we had found a Radisson Motel undergoing a character re-tread. We were surrounded by a phalanx of recently furnished motel rooms, with pools nestled among groves of bougainvillea and palm. It was a secretive, beautiful, and recently built paradise in the heart of an area that had been tidewater boondocks five decades earlier.

Cocoa Beach is a strange place; it is Rocket City USA. A handful of miles to the north is Cape Canaveral Air Force Base—as we have repeatedly seen, the real heart of Spaceport USA—while just beyond is the sprawling bulk of the Kennedy Space Center complex and Pad 39 from which the Apollo moon missions launched, from which the space shuttles still launch, and from which one day the new and wonderful *Orion* spacecraft will launch. In his extraordinary and lyrically written masterpiece *The Right Stuff*, Tom Wolfe points out the debt the American quest for space owes to Cocoa Beach. Until the late 1950s, Cocoa Beach was—to use his term—distinctly low-rent. It was the kind of place where American blue-collar workers with their plaid picnic baskets and Ban-Lon loafers went to vacation among the sweltering subtropical sun of the American Southeast, while the beach huts rotted gently in the dripping air and the no-see-um bugs performed arterial surgery on your lower legs.

And then something extraordinary happened; in the summer of 1958, NASA was born, and Route A1A—the characterless stretch of tarmac that runs from Miami Beach to Jacksonville—became the highway to the stars. Within a handful of years, the decaying beach shacks, the tired automobile dealerships, and the festering tidal swamps were replaced by brand-new burger joints and motels. It was here that the Mercury Seven came to unwind, and it was also here, in these unrecommended spots—where curiosity accounted for a little and loneliness accounted for a lot—that some of their more outré and amorous legends were born. Just as the space race experienced booms and busts, so did these ordinary mid-twentieth-century motels and catering joints rise and fall. Thus, it was with a feeling that was almost awe that Julie and I, with Jess still a babe in arms, first arrived in 1994 to discover the legends of Cocoa Beach for ourselves. We returned for more six years later with Susie, who had herself only recently arrived to swell the Corfield clan.

On our second visit, we found the Radisson quite unexpectedly packed with Boeing executives. To our delight, they were there because they had scheduled a rocket launch of one of their Delta V vehicles for the following day. Imagine! Quite by chance we had arrived at Spaceport USA the day before a rocket launch. We were beside ourselves with excitement. We discovered that this launch had been scheduled because an "issue" had arisen during the previous launch.

Reading between the lines, I soon learned that "issue" was Boeing slang for a rocket exploding midflight and spreading flaming debris hundreds of miles downrange across the Atlantic Ocean.

The next morning we were up and out early, finding the closest place to legally park near Pad–17 at Cape Canaveral Air Force Station. Minimum safe distance for the launch we were about to witness was three miles; for shuttle launches, it is eight. We stood on a beach in the stillness of the Floridian dawn, monitoring our watches and waiting for an event whose timing we had no way of estimating. Suddenly a white thunderbolt arced up from beyond the trees in an eerie and perfect silence. Seconds later the sound hit us, a continuous scream of combusting gases that swiftly crescendoed as the Delta V hurled itself at the sky. This business about rockets rising with implacable slowness as they gain momentum is nonsense. Within a few seconds, the Delta was a diminishing dot on the horizon, and the only record of its passage was a thick white vapor trail dissipating in the suppurating air.

## LOST HORIZONS

Six years later, my family and I waited for another rocket launch, sitting in the quiet of my study as, on the computer monitor in front of us, a different spacecraft prepared for a much longer journey than that taken by that Boeing Delta V. The *New Horizons* spacecraft blasted skyward from Launch Complex 41 on January 19, 2006, at the start of what would be the fastest transit of the solar system ever. After engine shutdown, the machine was already traveling at 36,600 mph, the fastest spacecraft ever built. As I write—less than a year after its launch—*New Horizons* is already within hailing distance of Jupiter, having passed through the asteroid belt without mishap. At Jupiter, it will pick up even more speed during a Minovitch-style gravity assist maneuver, then will fling itself into the depthless tracks of the solar system, eschewing the locations that its predecessors made famous—Saturn, Uranus, and Neptune—and heading straight to Pluto.

But *New Horizon*'s mission is touched with irony because, as it heads into the uncharted vastness of the outer solar system, Pluto, the planet it was sent to study, is no longer a planet.

## THE HUNT FOR PLANET X

In 1894 Percival Lowell, with his extensive personal fortune, set up an astronomical observatory in Flagstaff, Arizona, specifically to observe Mars. By the start of the twentieth century, however, Lowell had begun to observe Neptune and refine its orbital path. He concluded that its orbit—like the orbit of Uranus—was perturbed, and that the reason for this had to be the existence of another planet in a more distant orbit. Lowell predicted where this unknown planet should be, but his search drew a blank; he determined that he needed to refine his calculations. He did so and presented a paper—"Memoir on a Trans-Neptunian Planet"—to the National Academy of Sciences in the United States on January 13, 1915. He was so confident of his results that he wrote to his chief astronomer at Flagstaff, "I am giving my work before the Academy on January 13th. It would be thoughtful of you to announce discovery [of planet X] at the same time." Alas, he was disappointed; no discovery was made, and a humiliated and dejected Lowell wrote to his brother, "that X was not found was the sharpest disappointment of my life." Lowell died only a year later, having never found his planet X.

In 1929 a young astronomer named Clyde Tombaugh arrived at the Flagstaff Observatory and immediately applied himself to the decades-old problem of planet X. He used a new technique to search for the planet, which entailed comparing two negative photographic plates of a patch of sky taken some time apart and superimposing them. Objects on both plates—such as stars—would be superimposed on each other, but objects on one plate only—when a planet appeared in the field of view—would show up as separate dots. The device was called a "blink comparator" and made use of one of the most exquisitely sensitive detectors known to science—the human eye.

The approach finally paid off on the afternoon of February 18, 1930. Years later Tombaugh wrote, "I suddenly came across the image of Pluto! The experience was an intense thrill, because the nature of the object was apparent at first sight." Several more nights observing the new planet followed, and on March 13, 1930—the seventy-fifth anniversary of Lowell's birth—the discovery of planet X was officially announced.

Because the discovery was made at Lowell's observatory, the right to name the new planet belonged to the Flagstaff astronomers. Planet X's discoverer, Clyde Tombaugh, urged the observatory's director, Vesto Slipher, to suggest a name quickly to definitively establish the planet's identity. Vesto Slipher was a significant astronomer in his own right. He used spectroscopy to measure the rotational periods of planets and was the co-discoverer of galactic red shifts (the technique by which the speed of galactic movement is measured). Slipher had hired Tombaugh and supervised his work during the long hunt for planet X.

Suggestions for the new planet's name came from all over the world, but the most vocal contributor was Percival Lowell's widow, Constance. In a display of escalating hubris, she suggested Zeus, Lowell, and, finally, Constance. Her suggestions were greeted with a marked lack of enthusiasm. Other names based on mythology, such as Chronos and Minerva, were much more popularly received.

The suggestion to name planet X Pluto originated in a place and culture very far from the thin air of the Arizona mountains, in the bustling university town of Oxford in England. Venetia Burney was the eleven-year-old granddaughter of a former Bodley librarian named Falconer Madan. Venetia was passionately interested in classical mythology as well as astronomy, and she suggested that the

Roman equivalent of Hades might make a good name for the newly discovered object at the edge of the solar system. Her grandfather passed the suggestion on to Professor Herbert Hall Turner, the Savilian professor of astronomy at Oxford, who then cabled the suggestion to colleagues in America. Almost immediately the name was greeted with enthusiasm and was officially adopted on May 1, 1930.

By the time I was born in 1962, Pluto had come to serve as a reminder of the sheer size of the solar system. During my childhood, as I developed my lifelong interest in science, I knew Pluto as the distant speck that men of science had spent lifetimes trying to find. It was an iconic example of how pure science should be—the selfless quest for knowledge for its own sake—for what possible use is a tiny planet on the edge of forever? Pluto was the gatekeeper of the solar system—the last bastion of civilization before the howling winds of the interstellar wilderness begin.

Pluto does not have a significant atmosphere. It has a thin envelope of gases that are mostly nitrogen, methane, and carbon monoxide, with traces of ethane. When Pluto is farthest from the sun, the nitrogen and carbon monoxide form a super-cold snow. As Pluto moves toward perihelion, the solid nitrogen sublimates to gas, causing an "anti-greenhouse" effect. Much as evaporating sweat cools the skin, Pluto's sublimating nitrogen snow cools its surface. At best Pluto's atmosphere is unimpressive, being about 1/700,000 that found at sea level on Earth. Strangely for such a distant body, the surface of Pluto is remarkably varied. Maps from the Hubble Space Telescope show that the side of Pluto facing Charon has more methane ice, whereas the opposite hemisphere has more frozen nitrogen and carbon monoxide.

For the first several decades after its discovery Pluto was thought to be a reasonably sized planet—equivalent in mass to the Earth. But as data accumulated about our solar system's gatekeeper, it turned out that the mass estimate was inaccurate by a very large margin. In

1978 Pluto was discovered to possess a moon, Charon, that orbits in very close proximity to its primary and that had been included in estimates of Pluto's size. Subtracting Charon's mass leaves Pluto as not only the smallest and least massive of any planet in the solar system but, in fact, less massive than seven of the *moons* of the solar system: Ganymede, Titan, Callisto, Io, Europa, Triton, and our own moon. As we shall see, such diminutive stature did not help it when it came up for parole at the International Astronomical Union in the summer of 2006. It turned out that size matters very much indeed.

Another thing that did not help Pluto maintain its planet status was its weird orbit. To be taken seriously as a driver in most civilized countries of the world, you need a driver's license, which proves you know the highway code, can handle traffic lights, and can make it around gyratory systems without causing a pile-up. Similarly, the nomenclature of our solar system includes rules. Planets orbit the sun in a neat counterclockwise manner, all in one plane—the plane of the ecliptic. Not only is Pluto's orbit widely asymmetrical (for twenty years out of every two hundred it performs an illegal undertaking maneuver and slips closer to the sun than Neptune), but the plane of its orbit is highly inclined to the ecliptic. Pluto, it turns out, does not have a license.

Such anomalous behavior makes its bona fides very suspect. And, to make matters worse, it turned out that such maverick behavior was not limited to just Pluto.

## REBELS WITHOUT A CAUSE

Strangely enough, even after discovering Pluto, in an effort to square apparent gravitational anomalies, the hunt for a planet X continued. The indefatigable Clyde Tombaugh continued his search, in the process discovering asteroids, comets, and even a variable star. A U.S. Naval Observatory astronomer, Robert G. Harrington, continued the search right into the 1980s and 1990s, because his calculations indicated that there surely must be something besides Pluto responsible for perturbing the orbits of Uranus and Neptune. The deep-space probes *Pioneer 10* and *11* and *Voyagers 1* and *2* eventually provided the answer. As these probes skimmed past the outer planets, the change in their acceleration allowed them to accurately measure the gravitational pull—and hence the mass of the planets—very accurately. These measurements showed that the Earth-based estimates of the masses of several of the outer planets were inaccurate. When the correct masses were fed into the calculations, the orbital anomalies that had led people to hypothesize the existence of a planet X simply disappeared.

Only a few years later, in the early 1990s, advances in computerized telescopic techniques showed that the space beyond Pluto is very far from uninhabited, however. At first, just a few, but then dozens of subplanet-sized objects were detected by this method. The swarm of objects that infest this volume of space are now known as trans-Neptunian objects. They include any object in the solar system that orbits *on average* at a greater distance from the sun than Neptune. There are three main categories of trans-Neptunian objects: Kuiper Belt objects, Scattered Disk objects, and Oort Cloud objects.

The first scientist to suggest the existence of a belt of objects at the edge of the solar system was Frederick Leonard, who proposed the idea in 1930, the same year Pluto was discovered. The idea was revived by the astronomer Kenneth Edgeworth in 1943 and expanded in 1951 by the American astronomer Gerard Kuiper, who suggested that the belt might be a source of short-period comets. When an object actually was discovered orbiting in the region predicted by Kuiper, the belt was named after him. Our understanding of the Kuiper Belt has massively expanded since 1992, largely through advances in computer technology and telescopes fitted with image intensifiers.

Beyond the edge of the Kuiper Belt lies the region of so-called Scattered Disk objects. This region contains a population of icy minor bodies. The innermost portion of the Scattered Disk overlaps the Kuiper Belt, while its outer limits extend far from the sun and spread above and below the plane of the ecliptic. The formation of the Scattered Disk is still not fully understood, but many scientists believe it was formed when more conventional Kuiper Belt objects were scattered by gravitational effects from the outer planets, particularly Neptune. Many Scattered Disk objects move as far vertically as they do horizontally. An example, as discussed later, is Eris, one of the most recent of the Scattered Disk objects to have been sighted.

Centaurs are slightly different animals, icy planetoids that orbit between Jupiter and Neptune, continually crossing and re-crossing the orbits of the large gas and ice giants. They are probably Kuiper Belt objects that were sufficiently gravitationally perturbed to fall toward the sun rather than away from it and toward the Scattered Disk region.

The final class of trans-Neptunian objects is the Oort Cloud. This spherical cloud of comets is situated almost a quarter of the way to the nearest star system—Proxima Centauri. The Oort Cloud is thought to be the source of most, if not all, of the comets that flash periodically through the inner solar system. The concept of the Oort Cloud dates back to the heady days of Pluto's discovery; in 1932 the Estonian astronomer Ernst Opik proposed that comets originate in an orbiting cloud at the outer edge of the solar system. In 1950 the idea was expanded by the Dutch astronomer Jan Hendrick Oort, who noted that, since comets lose material each time they pass through the solar system, they must eventually be destroyed. Since comets continue to appear, Oort theorized that there must be a source where cometary material originates. It is now thought that the Oort Cloud contains millions of cometary nuclei that remain stable because the sun's radiation is too weak at that distance to evaporate them.

The Oort Cloud is thought to be the remnants of the original solar nebula from which the inner planets of the solar system coalesced 4.5 billion years ago. Astronomers believe that it initially formed much closer to the sun but was expelled from the solar system proper by the gravitational interactions of planets and asteroids—particularly Jupiter—and placed into highly elliptical, long-period orbits. This process also pushed the Oort Cloud out of the plane of the ecliptic, which explains its spherical shape.

It was Edmund Halley, Savilian professor of astronomy at Oxford in the early eighteenth century, who had one of the earliest insights into the nature of comets. Building on the work of his old friend and mentor, Isaac Newton, Halley applied Newton's laws to comets. In a 1695 letter to the Cambridge mathematician, Halley speculated that the comets of 1531, 1607, and 1682 were, in fact, successive appearances of the same comet. In 1705 Halley published the most important work of his life, *Synopsis of the Astronomy of Comets*, in which he gave the calculated orbits of twenty-four comets that were observed between 1337 and 1698. The work was groundbreaking, because it was the first time that comets had been treated as a separate but distinct part of our solar system. Halley predicted that the comet of 1531/1607/1682 would return in late 1758 or early 1759. His prediction was triumphantly confirmed, and the comet has been known ever since as Halley's Comet.

So what are comets? They are small bodies composed of a mixture of rock grains and condensed gases that typically have very large and elliptical orbits bringing them from beyond the orbit of Pluto to close to the sun. As they approach the sun, they warm, and the frozen gases thaw and surround the center of the comet—the nucleus—in a diffuse cloud called the coma. Usually this small, bright nucleus can be observed at the center of the comet. As they near the sun, comets develop long, luminous tails of gas that can extend for millions of kilometers. They are also accompanied by an envelope of hydrogen gas and another, invisible tail of ionized gases. Each time a comet approaches the sun, it loses some of its mass by evaporation; hence it eventually becomes just another chunk of rock. On a cosmological timescale, therefore, comets are some of the most short-lived objects in the solar system.

In 1985 Halley's Comet became more than a distant speck to the human race when it was visited by the ESA space probe *Giotto*. The probe came within six hundred kilometers of the comet, taking photos and analyzing its composition. Since then other comets have been visited, including Comet Grigg-Skjellerup (also by *Giotto*) and Comet Borrelly by the *Deep Space 1* probe in 2001.

## THE PLANET HUNTERS

Pluto's strange orbit and small size had been creating whispers about its classification in the astronomical community for decades. Its status might have gone unchallenged indefinitely, however, had it not been for the announcement in March 2004 that a tenth planet had been found. The planet was named Sedna after the Inuit goddess of fertility.

Sedna was discovered by noted planet hunter Michael Brown of Caltech, together with Chad Trujillo of the Gemini Observatory (with telescopes based in Hawaii and Chile) and David Rabinowitz of Yale, on November 14, 2003. It was found using Yale University's 160-megapixel camera and telescope instrument. Brown and his team estimate that Sedna has a diameter roughly three-quarters that of Pluto and that it has a highly elliptical orbit around the sun, with an orbital period of about 12,000 years.

Sedna is so far from the sun—roughly three times further away than Pluto—that its temperature never rises above about –240 degrees C. It is nearly as red as Mars and, unlike Pluto and Charon, has very little methane or water ice on its surface. Some speculate that Sedna's dark red color is caused by a globe-encircling hydrocarbon sludge of tholin. In this respect, it may be similar to Neptune's moon Triton.

Sedna was not the first body to be discovered orbiting beyond Pluto. In 1992 David Jewitt and Jane Luu of the Mauna Kea Observatory in

Hawaii discovered the first trans-Neptunian object. Its International Astronomical Union classification is (15760) 1992 QB1. Curiously, the object has never been given a proper name, although an entire category of trans-Neptunian objects has been derived from its designation—cubewanos ("Q B one oh"). Technically, cubewanos are Kuiper Belt objects that orbit beyond Neptune and are not controlled by an orbital resonance with that planet.

In late 2000, Varuna was discovered orbiting beyond Pluto by Spacewatch scientist Robert McMillan, although it had apparently been noted as early as 1953. In 2002 Quaoar was discovered by Brown and his team. Quaoar is the name of a creation deity of the Native American Tongva people, who lived in the Los Angeles area. When it was found, its diameter of slightly more than 1,200 kilometers made Quaoar the largest-known body beyond Pluto's orbit. It was the first trans-Neptunian object to be measured directly by the Hubble Space Telescope.

Sedna's size—very nearly that of Pluto—cast doubt on Pluto's status as a planet, but in 2005 came the coup de grâce. Once again, Mike Brown was the discoverer of an object orbiting beyond Pluto, but this time there was a vital difference. Object UB313—as it was initially known—was *larger* than Pluto, and Brown and his co-workers nicknamed it Xena and started referring to it as the "tenth planet," the uncertain status of Sedna notwithstanding. What was more, the new planet orbits *beyond the edge* of the Kuiper Belt; it is thus a Scattered Disk object.

People had had enough. It was time to sort things out. The International Astronomical Union (IAU) put together a team of nineteen experts to decide on the definition of a planet. In August 2006, the issue was put to a vote of delegates at the union's meeting in Prague. The delegates voted that, to be considered a planet, an object must fulfill three criteria. First, it must be big enough for its gravity to have made it a round object; second, it must orbit the sun and not be a satellite of another body; and third, it must be tidy enough to have cleared other bits of debris from its orbit. Pluto, a member of the untidy hodgepodge of trans-Neptunian objects, failed all three criteria and was summarily demoted to the new category "dwarf planet."

For the mission scientists of the *New Horizons*, it was a bitter blow. Their craft was no longer going to investigate a planet. Alan Stern, *New Horizon*'s principal investigator, was incandescent. And he was not alone—many members of the IAU thought that the three-part definition was overly prescriptive.

The debate brought to my mind memories of the way biology has developed in the past few decades. When I was an undergraduate, a major debate was whether the classification of plants and animals should serve only a pigeon-holing purpose or whether it should reflect the underlying principles that dictate the diversity of life on Earth—which is to say, evolution. The difference is not trivial and determines whether we simply pigeon-hole species for convenience or develop a scheme that recognizes *why* they are different. The debate over the definition of a planet seems to me to be precisely equivalent, and it seems eminently sensible that Pluto should not be considered a planet but rather a member of the much larger group of trans-Neptunian objects. After all, the eight other planets were apparently formed individually and uniquely out of the solar nebula 4.5 billion years ago (although I do harbor some heretical reservations about Neptune), whereas Pluto is simply one member of a much larger community—trans-Neptunian objects—that have their own identity and history. The crucial point is that the reclassification of Pluto reflects the fact that Pluto was not formed *in the same way* as were the other planets.

Only a month after the Prague vote, Xena shed her nickname when the IAU formally accepted Brown's suggestion of Eris as the name for the planetary body—which by this time had been discovered to have a moon. The moon was named Dysnomia. The names have a certain delicious irony for, in Greek mythology, Eris caused a fight over a golden apple, which led to the Trojan War, and Dysnomia is the goddess of lawlessness. Both names echo the battles that have raged and continue to rage over Pluto's planetary status.

## REVELATION SPACE

Perhaps the most extraordinary revelation in planetary science has been the discovery that other planets orbit other stars. Since the first golden age of planetary astronomy in the seventeenth and eighteenth centuries, it has been assumed that other star systems must have planets (as well as the derivative idea that there may be life around other stars). Only in the past two decades, however, have we obtained proof of extrasolar planets.

Incredibly, claims of extrasolar planets date back to the time of Darwin. In 1855 Captain W. S. Jacob of the East India Company, working at the Madras Observatory, reported orbital anomalies in the binary star system 70 Ophiuchi. He suggested that these orbital anomalies were a "highly probable" indication of a planetary body in that star system.

70 Ophiuchi is a binary star system only 16.6 light years away in the constellation Ophiuchus. The primary is an orange main sequence star, and the secondary is an orange main sequence dwarf. The constellation Ophiuchus was one of the forty-eight constellations known to Ptolemy and one of the thirteen constellations of the zodiac; it is the only one not associated with an astrological sign. It is commonly depicted as a man supporting a serpent and is well known to followers of Frank Herbert's epic *Dune* series as the constellation that contains the star 36 Ophiuchi, about which orbits Giedi Prime, the ravaged home planet of the vicious Harkonnen family who are the arch rivals of the series' hero, Paul Atreides. 70 Ophiuchi also makes an appearance in *Dune* as the star system about which the planet Sikun orbits.

In the 1890s, Jacob's idea was taken up and championed by Thomas Jefferson Jackson See, one of the strangest characters of late nineteenth-century and early twentieth-century astronomy. See, born near Montgomery City, Missouri, on February 19, 1866, was the sixth of nine children sired by Noah See, a local, well-to-do farmer. His early education was typical of country boys in the years following the Civil War: four months of schooling a year—while the snow and ice of a Midwestern winter covered the fields—in a tiny schoolhouse. The rest of the year, See was required to work on the land. When he was seventeen, it became clear that he was a potentially brilliant scientist, and See was allowed to attend a proper high school in the county seat.

In school, See so distinguished himself in the physical sciences that his father was persuaded to send him to the state university in Columbia. Though he continued to excel in all the sciences, astronomy attracted him the most. He graduated in 1889 as class valedictorian. While doing postgraduate work at the University of Berlin using the Royal Observatory's nine-inch refractor, See developed an interest in double stars that dominated most of his young professional life and that would also sow the seeds of his professional destruction. Oblivious to the turbulent waters ahead, however, See received his doctorate from the University of Berlin on December 18, 1892.

Following his graduate work, See was appointed to the faculty of the new University of Chicago under the brilliant astrophysicist George Ellery Hale (later of comet fame). It was a time of enormous excitement and expansion at Chicago for Hale. The university's president, William R. Harper, had recently persuaded Charles T. Yerkes—the Chicago streetcar tycoon—to purchase forty-inch glass lenses from the Massachusetts shop of Alvan Clark & Sons. The huge lenses were destined to become the optical heart of a telescope that would be the centerpiece of the new Yerkes Observatory, which would become one of the most famous astronomical observatories in the United States.

In Chicago, See was assigned the task of cataloging double stars. Logging and refining the information on the forty double star systems then known was a logical extension of his graduate work, and the young professor tackled the problem with gusto. See was also well aware that it was a high-profile project, and he quickly realized that, if handled correctly, it would provide an immeasurable boost to his reputation. Like many young academics before and since, See was hugely ambitious, and he craved not only the recognition of his peers but also the recognition of the wider world. He began submitting papers almost monthly to the prestigious *Astronomical Journal.* He also maintained and developed the connections made during his time in Germany by making regular submissions to *Astronomische Nachrichten*, the premier German astronomical journal. This activity was also the start of a prolific career as a popular science writer, for he also began sending articles regularly to scientific monthlies such as the hugely influential magazine *Popular Astronomy*.

In 1895 Sea submitted a letter to the *Astronomical Journal* on the orbit of 70 Ophiuchi in which, like Jacob before him, he noted the presence of orbital irregularities. But only four years later, in 1899, one of See's own graduate students, Forest Ray Morton, published a paper showing that 70 Ophiuchi's postulated dark satellite could not exist. His mathematical proof showed that the orbit of such a satellite would be highly unstable. See wrote a reply that was self-serving, arrogant, and abusive, and although the *Astronomical Journal* only published an excerpt, it was enough to persuade the editors that they should henceforth effectively ban See from submitting to their magazine. Although See lived to be ninety-six, he never again published in the *Astronomical Journal.* The 70 Ophiuchi debate was sufficiently vicious that it permanently blighted See's subsequent career and shelved the notion of extrasolar planets for several decades.

Given this story, it is doubly ironic that, some six decades later, the next star to be suspected of having a planetary companion should be in the constellation Ophiuchus. Peter van de Kamp arrived at the University of Virginia's McCormick Observatory from his native Holland on March 1, 1923. After a year, he moved to the famous

Lick Observatory in California, where he received his doctorate in 1925. He returned to Virginia after completing his graduate work, and he stayed there until leaving in the spring of 1937 to take over as the director of Swarthmore College's Sproul Observatory. At Swarthmore, van de Kamp started the observations that would make both him and the fourth nearest star to our sun famous. Van de Kamp was an assiduous observer and spent years studying Barnard's Star—an otherwise unprepossessing red dwarf star whose only claim to fame is its possession of the largest proper motion of any star relative to the sun. In fact, Barnard's Star is approaching our sun so rapidly that, within 10,000 years, it will be closer to us than any of the three stars that comprise the Alpha Centauri system.

For twenty-five years, van de Kamp examined the motion of Barnard's Star, often asking colleagues and collaborators to check his photographic plates to eliminate any hint of individual bias. He was a most meticulous observer, but the job was difficult—he and his collaborators were looking at the tiny image of a distant star on a photographic plate, searching for displacements of only a micron. By 1963, however, van de Kamp was sure that the motion of the star showed a wobble indicating the presence of an unseen planetary companion orbiting the star in a highly eccentric orbit. To perturb even a small star like Barnard's Star requires a relatively large planetary companion, and van de Kamp concluded that it must have a planetary mass similar to that of Jupiter and be orbiting very close to the star.

Unlike See, van de Kamp was a highly respected astronomer and scientist, and for almost a decade the astronomical community was inclined to believe him. But in 1973, two papers were published that undermined van de Kamp's claim. George Gatewood was a doctoral student at the University of Pittsburgh under the direction of the noted astrometrist Heinrich Eichhorn. Although Gatewood was not particularly interested in Barnard's Star, he was persuaded to study it and did so using two different telescopes, a new automated photographic plate measurement device, and his supervisor's data reduction algorithm. They could find no evidence for the orbital perturbation van de Kamp claimed. That same year, John Hershey, a graduate student at van de Kamp's own observatory, pointed out that Gliese 793, a star he was studying, showed exactly the same orbital perturbations that van de Kamp had noted. The conclusion was hard to avoid—some kind of systematic bias was being introduced into the data being collected by the Sproul Observatory.

But what could it be? It was Hershey, going back through the maintenance logs, who found the problems that had resulted in the phantom perturbations: a change to the cast iron casing that held the lens of the telescope, together with a change in the photographic emulsions employed in 1949 and then in 1957. Both years were the years when van de Kamp had noticed the orbital perturbation that had led to the claims for Barnard's Star. The implication was immediate and indisputable: Barnard's planet was nonexistent. In the language of science, it was an artifact.

For the rest of his life, Peter van de Kamp refused to acknowledge his error. He continued to affirm the existence of not one but two planets around Barnard's Star as late as 1982. Like Jefferson See before him, van de Kamp's career suffered from his infatuation with extrasolar planets. Although for most of his career he had been an outgoing and admired man, toward the end of his life he became increasingly embittered. Partly this bitterness was due to the criticisms coming from so close to home—not only had John Hershey challenged an idea that had originated at the Sproul Observatory but, in 1976, van de Kamp's successor, Wulff Heintz, began to publish papers refuting the existence of Barnard's planet (which van de Kamp had increasingly taken to calling "van de Kamp's planet"). After a while, the former director barely spoke to his successor. For the rest of the 1980s and 1990s, as more and more data accumulated with the aid of progressively better technologies, it became clear that there was no sign of even one planet orbiting Barnard's Star. The ultimate irony is that, so certain are astronomers today that Barnard's Star has no planetary companions, it is now used to debug the Hubble Space Telescope. If the Hubble's operators discern a wobble while imaging Barnard's Star, they immediately reach for their screwdrivers.

The controversy over Barnard's Star—like that of 70 Ophiuchi before it—poured cold water on the search for extrasolar planets for another twenty years. As with the discovery of dwarf planets, however, the search for extrasolar planets was immeasurably helped by advances in technology. In 1988 Canadian astronomers Bruce Campbell, G. A. H. Walker, and S. Yang made observations that suggested that there are planets orbiting the star Alrai (also known as Gamma Cephei). Their measurements were at the limit of what was possible at the time, and not until 2003 did improved techniques allow the planets' existence to be confirmed.

In 1992 radio astronomers Aleksander Wolszczan and Dale Frail announced the discovery of planets orbiting the pulsar PSR 1257+12. It was not long until the discovery was confirmed, making it the first definitive sighting of an extrasolar planet. It was one of the most significant moments in the history of astronomy. Both PSR 1257+12 and Alrai are pulsars, the remnants of a supernova, and it is thought that the planets that orbit them were also formed by that cataclysm. The question remained, would conventional planets be found?

That question was answered when, on October 6, 1995, Michel Mayor and Didier Queloz of the University of Geneva, working at an observatory in the south of France, announced their discovery of a planet orbiting 51 Pegasi, an ordinary main sequence star such as we met in Chapter 1. It was the finding that opened the floodgates of extrasolar planetary discovery. To date, over two hundred extrasolar planets have been found, and more are being located almost daily. Advances in high-resolution spectroscopy, which allow extrasolar planets to be detected indirectly by measuring their gravitational influence on their parent stars, are primarily responsible for this breakthrough. In addition, several extrasolar planets have been detected by looking for variations in a star's apparent luminosity as a planet passes in front of it.

In fact, since the pioneering work of Mayor and Queloz, many other exoplanets have been discovered by teams working at the University of California's Lick and Keck observatories. Many of the early detections of the 1980s have been confirmed, and at least twenty star systems are now known to have more than one extrasolar planet. Extrasolar comets are known to be common, too.

Extrasolar planets are difficult to detect, however. Compared to their parent stars, they are usually less than a millionth of their parent star's brightness. It is like looking for a candle beside a million-watt searchlight. For this reason, most of the exoplanets discovered so far have been large (usually larger than Jupiter). The main detection methods are astrometry, the Doppler method, pulsar timing, the transit method, gravitational microlensing, and using circumstellar disks.

Using the astrometry technique, a star's movement is carefully measured. The gravitational influence of a planet will cause the star to wobble slightly around their common center of mass. The Doppler method consists of measuring a star's spectrum and noting any variations that could be due to a wobble imparted by an orbiting planet. This method is the most effective technique used by extrasolar planet hunters. The pulsar timing method relies on the fact that a pulsar emits regular radio waves as it rotates. Slight variations in the timing of these pulses can be used to infer gravitational perturbation caused by the presence of an extrasolar planet. The transit method relies on the small drop in a star's luminosity when an orbiting planet passes in front of it. Gravitational microlensing is when a star's gravitational field behaves like a lens, magnifying the light of a background star. If the lensing star has a planet, then the planet's own gravitational field can alter the lensing effect. Finally, many stars are surrounded by circumstellar disks of dust that can be detected because they absorb ordinary starlight and re-emit it as infrared radiation. Concentrations of infrared radiation can be used to infer the presence of planets.

The hunt for planets outside our own solar system remains one of the most exciting aspects of modern astronomy, and several space missions are already planned that will use these methods for detecting extrasolar planets. Space-based observations have the advantage of avoiding the distorting influence of the Earth's atmosphere, plus a wider range of infrared wavelengths can be employed. The ultimate aim of many of these projects is to detect planets of similar size and composition to the Earth. The impetus to understand the heavens and discover whether or not we are alone in them continues unabated.

Since the discovery of the planets orbiting 51 Pegasi, the rules concerning naming extrasolar planets have been simplified and standardized. The first planet to be discovered in a given star system is given the letter "b" ("a" refers to the star), and subsequent planets, as they are found, are given the next letter in the alphabet. Thus the first planet discovered in the 51 Pegasi system was given the name 51 Pegasi b, the next, 51 Pegasi c, and so on. Letters refer to the order of discovery, not the distance from the star. The detection in the 1980s of the planets orbiting PSR 1257 predates the discovery of 51 Pegasi and so use an older system of nomenclature. It is common for some extrasolar planets to have unofficial nicknames. Thus HD 209458 b is unofficially called "Osiris," and 51 Pegasi b is called "Bellerophon," a nod to the classic 1950s movie, *Forbidden Planet.*

Most of the exoplanets found to date orbit stars similar to our own sun, that is, main sequence stars of spectral categories F, G, or K. One reason for this is simply that we have tended to focus on such stars in the hope that we will find planets similar to our own Earth. Stars that contain a higher concentration of metals along with their hydrogen and helium are more likely to have planets. Most exoplanets discovered to date have high masses; indeed, many are more massive than our own Jupiter. This probably simply reflects the fact that more massive planets are easier to detect than smaller planets, which further underlines the need for space-based observation platforms. It is highly likely that lower-mass planets will prove to be more common than higher-mass ones. Consider our own solar system—the vast majority of objects in it are small; only Jupiter and Saturn are high-mass planets.

Although many extrasolar planets are probably similar to Jupiter and Saturn (gas giants), it is likely that the few smaller examples found to date are rocky. As detection techniques improve, we expect that many more of these smaller planets will be found, thus improving the prospects for locating worlds similar to Earth.

The exoplanets found to date tend to orbit closer to their parent star than any of the planets in our own solar system, and it seems likely that most exoplanetary systems have only one or two very large planets. If so, this finding could have profound implications for the hunt for extraterrestrial life, because it implies that star systems like our own are rare. Many exoplanets have highly elliptical orbits, which remains a puzzle because our current theories of planetary formation suggest that most should have more circular orbits. It is a further indication that our own solar system may be unusual, since all of its planets have orbits that are more circular than elliptical. Several of the extrasolar planets discovered so far do orbit within their parent star's habitable zone, but all are giant planets more similar to Jupiter than to Earth. However, these planets could have moons, which might be the abode of life as is suspected for Europa and perhaps Titan in our own solar system. One of the most exciting discoveries to date has been the detection by astronomers using the Hubble Space Telescope of an atmosphere on planet 2001, HD 209458 b. If atmospheres are common on extrasolar planets, the likelihood of life in other star systems takes a quantum leap forward.

The oldest extrasolar planet known to date lies in the globular star cluster M4, about 5,600 light years from Earth in the constellation Scorpius. It is the only planet known to orbit a stellar binary; one of the stars is a pulsar, and the other, a white dwarf. The planet is twice the mass of Jupiter and is estimated to be a staggering 13 billion years old, or almost the age of the universe. Two planets have been discovered that are probably rocky and that orbit conventional main sequence stars; they are 2004 Mu Arae d and Gliese 876 d.

## STRANGERS IN STRANGE LANDS

We live in the most extraordinary age of astronomical discovery. We have found that the worlds of our own solar system are much more diverse than ever imagined, and we may yet detect life, perhaps on one of the moons of giant Jupiter or Saturn. Our discovery of extrasolar planets continues apace, and we cannot be far from confirming the existence of rocky worlds similar to our own around other stars.

The next time you glance at the night sky, spare a thought for that incredible diversity and revel in it. One day in the not too distant future, our children will venture in person across the solar system rather than send just robot proxies, and, ultimately, they will travel beyond it, celebrating that diversity for themselves. And when that distant day dawns, our descendants will look back and remember our generation, because despite all the faults and petty squabbling with which we sallied forth to sully the latter years of the twentieth century, we will have at least given them the ultimate gift of all: a life for the stars.

# ACKNOWLEDGMENTS

Many thanks to Bill Frucht, Courtney Miller, and Jodi Marchowsky at Basic Books, Jane McGraw at Word Craft and to my agents Peter Robinson and Jill Grinberg for making the project happen.

Thanks to Bob Spicer, Phil Potts, Simon Kelley, and John Zarnecki for making me so welcome at the Open University. Other colleagues there whose support has been invaluable include Anthony Cohen, Nick Rogers, Mike Widdowson, Ian Wright, Iain Gilmour, Mabs Gilmour, and Colin Pillinger.

Librarians have, as always, been essential so thanks to John Hillsdon, Isabel McMann, and Jenny Colls at Oxford as well as all the staff at the Open University Library and Witney Public Library. Many friends in the radio, TV, and publishing world continue to support my work and writing so thanks in particular to Steve Wilkinson, David Sington, Melvyn Bragg, Charlie Taylor, James Cook, John Jarrold, and Mark Smalley.

Many good friends have supported me through the writing of this book, in particular Tom Duggan, Brian and Heather Boyt, Colin Holmes, Peter Kent, Mike Dean, Mike Durkin, Steven Eeley, Brian Heath, David Jackson, Terry Walsh, Ian Crook, Dennis and Christine Armstrong, Jon and Jo Oldham, Marvina George, Alan Edwards, Brian Guntrip, Carl and Claire Shuker, Chris and Dolores

Cox, Stuart McMath, Richard and Deborah Balding, John and Jayne Laurie, John Price, Neil Wynn, Frederic Doussin, Sheelagh Herrin, and my wonderful, supportive father, George Corfield.

Thanks also to the Society of Authors for support.

Particular thanks to Christopher Riley, David Hancox, David Rothery, and John Zarnecki for reading portions—or, in some cases, all—of the manuscript. Any mistakes or omissions of course, remain my own.

The title of Chapter 7 is borrowed from one of the extraordinary opus of songs by Pete Atkin and Clive James whose music has comforted and inspired me in equal measure for as long as my interest in space exploration.

First, last, and always, thanks and love to Julie, Jessica, and Susannah who share my enthusiasm for books, writing, and outer space, and who make it all worthwhile for me.

# INDEX